To Mayor Herent

BROKEN RIVER

Sam Penny
10-8-04

What Readers say about The 7.9 Scenario series:

Dr. David Stewart says, "Memphis 7.9, the first novel in The 7.9 Scenario series, is a gripping novel I couldn't put down. The characters, who seem real, are fiction, but the account, which may seem like fiction, is all too real. Based on solid science and the projections of many seismologists, Sam Penny has created the scenario of a major earthquake on the New Madrid Fault and its impact on the Midwest and its people. Sam's earthquake will happen some day and it could be in our lifetime. I can't think of a better way to prepare for it than to read and experience Sam's excellent, well-written accounts. They could save your life. Memphis 7.9 is part one of a series. I look forward to Sam's next book and will read it with interest."

Jim W says, "Sam Penny has taken a different approach to spreading the word about the New Madrid Seismic Zone and its associated risk. After more than fifteen years of studying available scientific and engineering literature concerning the New Madrid Seismic Zone, Mr. Penny decided against writing another scientific treatise, mainly for the reason that most people can't identify with them. Instead he decided to approach the issue as a fictional account based on information he had garnered from researched data from the Federal Emergency Management Agency and the U.S. Geological Survey. The 7.9 Scenario novels pull the reader in and capture the imagination while adding an element of "what if" that raises awareness to the hazard faced in the central United States."

Trent F says, "The 7.9 Scenario should be a wake up call for middle America. It tells the story of the earthquake from the standpoint of the impact on people, property, and America's economy. Written from the perspective of citizens, seismologists, politicians, and others impacted by the event, author Sam Penny strives to point out the need for preparation, by giving an overview of the potentially devastating societal and economic impact.

BROKEN RIVER

A Novel

Book 2 of The 7.9 Scenario

SAM PENNY

TwoPenny Publications

First TwoPenny Edition
1 2 3 4 5 6 7 8 9 10

ISBN 0-9755671-1-X

Published by TwoPenny Publications

205 Rainbow Drive #10503
Livingston, Texas, 7739-2005
www.the79scenario.com
TwoPennyPubs@the79scenario.com

Library Of Congress Number: 2004096551
Printed in the United States of America

Dedicated to
Wilma Jean Pexton Penny
1911-1998

My mother, who always wanted me to write.

ACKNOWLEDGEMENTS

I am blessed with a wife who is supportive, is good-natured, and is a good editor and critic. Without her presence, this book would never have been written.

Access to good technical information was a vital element in producing this book. Materials provided by the Central United States Earthquake Consortium, the Army Corps of Engineers, and Dr. David Stewart proved most helpful. I also want to thank Dave Ellerbrake for his advice on the steering and manning of large boats.

For quality assurance, I thank Stephanie Bernhagen and Marianna Nelson. Both are very fine editors who eliminated much of my archaic spelling, corrected my sentence structure, and improved my writing style. Since I had the final say, what remains is my responsibility.

PREFACE

The 7.9 Scenario really began October 17, 1989, when I saw the remains of the Cypress Freeway following the Loma Prieta earthquake in the San Francisco Bay area. A solid concrete structure, only a mile from where my company once had offices and that I had driven day in and day out, had simply crumbled. Why?

Over the next ten years as I read the studies into that question and about the chances for more devastating earthquakes in the United States, I came to understand that the danger of earthquakes is widespread. I learned to my surprise that the central United States had the potential for an earthquake ten times more damaging than anything we could have in California. I also kept pace with the advances in seismology and understanding of the mechanics of an earthquake.

The 7.9 Scenario took physical form the day I purchased a copy of "Damages & Losses From Future New Madrid Earthquakes" by Dr. David Stewart, Ph.D., written while he was Director of the Center for Earthquake Studies, Southeast Missouri State University, Cape Girardeau, Missouri. Developed for the Federal Emergency Management Agency, this document provided data for the estimated shaking intensity on a county-by-county basis for the entire central and eastern United States. It also provided the tools for county officials to estimate casualties and damage in their county for various sizes of earthquakes.

I searched the Internet for a summation over all affected counties from a New Madrid earthquake, but to my surprise I found nothing. It was as if no one wanted to look at the overall picture, only at the local effects of a big earthquake. But I wanted to know, so I created a large spreadsheet, a line for each county, and summed it all to produce a total casualty estimate for the country. The numbers were staggering: tens of thousands killed, hundreds of thousands injured, millions homeless, damage in the tens of billions of dollars. Why was no one pointing out the danger?

When I retired in 1998, I began studying my scenario calculations in earnest, searching for the most vulnerable points of failure in the cities and structures that lined the Mississippi and Ohio Rivers. I refined my models and spreadsheets to provide more control for asking "what if?" questions for various sizes and placements of earthquakes. I kept waiting for the USGS or FEMA to produce a definitive report calling attention to the level of danger that I sensed.

Shortly thereafter, I decided I would write a novel to shine a light on the danger. I felt writing a scientific treatise on the subject would be a waste of time, mainly for the reason that most people don't identify with such works. Instead, I decided to approach the issue as a fictional account based on information I had garnered from my research.

In 2003 I published "Memphis 7.9," Book 1 of The 7.9 Scenario. I used scenes left on the cutting room floor from the first book as the basis for "Broken River," Book 2 of The 7.9 Scenario. The two are companion novels, though one does not require the other.

It is my intention to write Book 3 with the recovery efforts in Memphis as a backdrop for the plot.

Our country cannot prevent the next giant earthquake on the New Madrid Fault, but we can plan for how to handle the disaster. Response to the saying, "if you fail to plan, you plan to fail," will determine how close The 7.9 Scenario numbers come to reality.

Sam Penny
August 29, 2004

and now

Broken River

**What happens when,
not if, an earthquake strikes
on the New Madrid Fault beneath the
Mississippi River?**

Prologue

Saturday morning at 4:58 Central Daylight Time Chris Nelson's computer workstation at the University of Memphis Seismic Laboratory received notice over the network of a 2.8 magnitude earthquake beneath the tiny town of Cooter in the southeast Missouri Bootheel. The system launched the program called Nelson, the foundation of Chris's earthquake prediction theory.

In 1811 and 1812 a series of great earthquakes struck the New Madrid Fault. Log cabins, lean-tos and a few tents comprised the man-made structures along the Ohio and Mississippi Rivers, home to maybe 5,000 white settlers and black slaves. Another 20,000 natives roamed the nearby wooded floodplains and rolling hills. The rivers flowed wild and free as log-boats and the first Mississippi steamboat plied the current near the bluffs that would become Memphis.

Today 55,000,000 souls of all extractions live in the nine states that shook the hardest nearly 200 years ago. Today these two mighty rivers and their major tributaries are man-made structures, reformed by the Army Corps of Engineers, the Tennessee Valley Authority, and the local Levee Authorities. Dikes line their banks, weirs control their flow, locks manage their traffic, and huge dams form lakes where valleys once prospered. Now the rivers form a giant power generation, transportation, and flood control system, a man-made structure destined to be broken.

Data streamed into the workstation over the phone lines from the myriad of remote sensors scattered along the New Madrid Fault. Interpretations arrived from Universities in Little Rock, St. Louis, and Jackson over the

Internet. Nelson used all the available information and executed Chris's latest Prediction Model, directing that the results be stored to the hard disk.

The analysis took three minutes to complete. Nelson stored its findings and returned the workstation to standby. Before the screen saver covered it over, the display proclaimed

```
Nelson Connected-Asperity Model, Case 141.
Prognosis based upon latest actiity at 04:58.
Epicenter on asperity #1 at 89.85W, 36.05N,
secondary on asperity #2 at 89.97W, 35.89N;
initiate at 09:33 CDT Saturday; magnitude 7.9
```

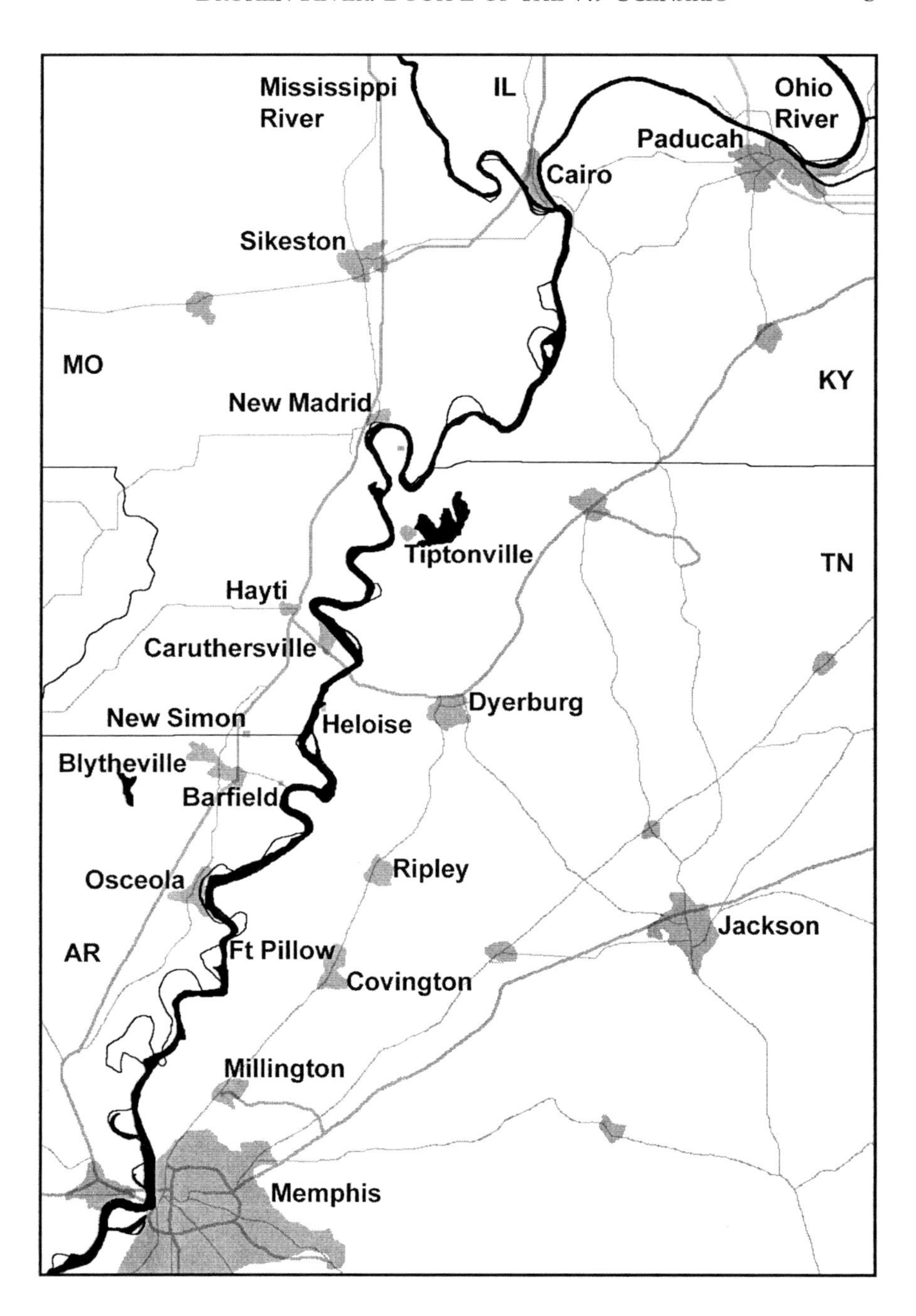

The Mississippi and Ohio Rivers

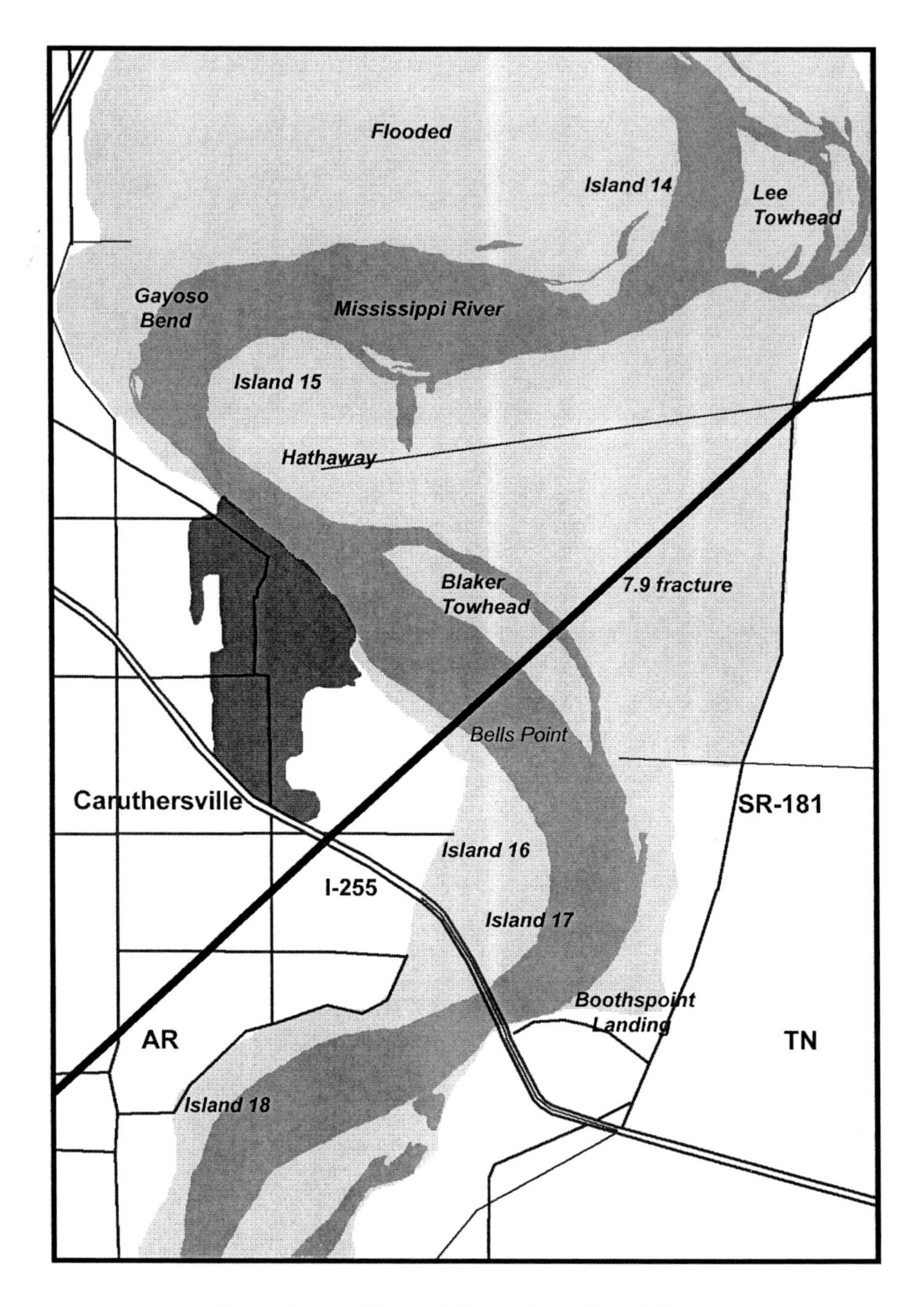

Caruthersville and Boothspoint Bridge

— 1 —

Approaching Caruthersville

Captain Buddy Joe Simpson rubbed the thick gold wedding band across his short curly red beard and listened to the ship's clock chime two bells. A few weeks ago he had adjusted the brass clock screwed to the wall to Central Daylight Time. The hands pointed to nine o'clock local time, Saturday morning.

He rotated his aching shoulder as he sat on the wheel house stool, his right hand resting on the rudder control of the 55-foot towboat, the *Lady Bird Jamison*. The reddish-gray hair atop his six-five frame brushed the low wheel house ceiling. He scanned the river ahead, looking over his tow of 12 covered grain barges loaded with soybeans and corn. Arranged three abreast by four long, the barges were bound downriver for Krueger Chemicals on President's Island just below Memphis.

His combined vessel measured 435 feet in length with a 105-foot beam. He sensed, more than heard, the hypnotic vibrations in the deck from the twin 1,895-horsepower diesel engines nestled in the hull below, pushing the tow through the water at eight knots relative, thirteen knots absolute by the GPS reading. With the Mississippi River running above flood stage, shipping channel currents ranged from four to seven knots. The boat had made good time coming past the Lee towhead, just west of Ridgely, Tennessee, and now ran two hours ahead of schedule.

He glanced at the solid green wall of trees trailing their leaves in the chocolate-colored water along the bank 200 yards starboard, then at the thermometer hanging in the shade by the open door. He murmured to himself, "It's 78 degrees already. Going to be another hot one. The humidity must be 90 percent. Sure looks like rain later on." He wiped his perspiring face and neck with the stained red bandana he kept in his flannel shirt pocket.

Using a dull stub of a pencil, Buddy Joe noted his location, speed, and heading in the log and reached for the handheld microphone of the marine radio. Selecting the frequency he had labeled "Lady Bugs," he thumbed the transmit button and spoke.

"This is Captain Buddy Joe Simpson of the *Lady Bird Jamison*, calling for the nine o'clock radio check of the Lady Bug Fleet. Hey, Paul? Denny? You guys, wake up. Dennis, report first." He released the button and waited for his fellow skippers on the *Lady Jane Wilson* and *Lady Janet Quayle* to respond.

The radio speakers came alive. "Captain Dennis Bugler of the *Lady Janet Quayle* checking in. I was expecting you, Buddy Joe. I'm pushing my tow of grain barges, three by four minus one, three miles above Cape Girardeau heading downstream to drop them off for you at the usual place below Cairo. We're still getting a little thunder around here from the storms left over from last night, but the weather's mostly moved northeast into Illinois. It should clear for the rest of the day. Over."

"Good hearing from you, Denny. I copy that. You'll have my next tow waiting for me at the Cairo buoy." He paused a moment. "Paul, are you awake for a change?"

"Captain Paul Taylor of the *Lady Jane Wilson* checking in. I'll have you know I've been working since five this morning. I just docked my tow at the McLaughlin Steel Works here in Osceola Harbor and have to finish the paper work. We have a day's layover before we pick up some empty barges and deadhead back to pick up another tow of steel from Paducah. Once we're docked, I thought I'd take the crew out to lunch. It's Ben's birthday. Would you believe that old coot's 50 years old today? He's been a great first mate. Thanks for recommending him to me, Buddy Joe. Over."

"Wish Ben happy birthday for me, will you? As for the *Lady Bird Jamison* we're rounding the bend into Caruthersville bound for President's Island with 12 full barges of grain. Should get there about eight o'clock this evening to drop off this load. I plan for us to take a day off–that is if the company doesn't find us another short job–and I'll watch NASCAR on TV tomorrow before heading back up to Cairo on

Monday." He rubbed his eyes and steered the tow three degrees starboard to line up on the next channel marker buoy.

"Guess that's it, fellows. And remember to watch your wake wash. Don't get the Army Corps on our necks like last week. Next radio check at eight bells. *Lady Bird* clear." He switched the radio back to scan mode and hung the hand-held microphone on its hook.

Chris Nelson sat on the vinyl-covered stool at the slick worn counter of the New Simon Truck Stop Café. He stirred the remains in his half-filled coffee cup and worried about his latest earthquake prediction. When he checked last night, what had begun a week ago as a prediction for a moderate magnitude 4.5 temblor on Sunday had escalated to a dangerous 6.1 estimated to initiate at 3:45 this afternoon. The clock to the side of the stove in the kitchen read 9:16.

Earlier that morning, he had squirmed from the sleeping bag next to his seismic monitoring station in the power substation lot west of I-55. He scratched his unruly mop of dirty blond hair and stepped into his sandals and shorts.

Then he called his assistant's voice mail and left a message, "Jenny, I'm driving over to Nasty's truck stop for some coffee and something to eat. When I checked last night Nelson said that the earthquake isn't due until three this afternoon so we have plenty of time to do the final check on the equipment. I'll call the University Seismic Lab some time after nine-thirty. Okay?"

For two years Chris, a graduate student in seismology, had refined his seismic prediction computer model for the New Madrid Seismic Zone, an area running from 45 miles northwest of Memphis to Cairo, Illinois, at the junction of the Mississippi and Ohio Rivers. His model had predicted with success two of the earthquakes along the fault in the past three weeks, and it now forecast a new temblor beneath New Simon on the Arkansas-Missouri border.

As more precursor events occurred throughout the week the projected magnitude of the predicted temblor grew and the estimated initiate time moved closer. The increase worried Chris. He had rationalized to Tina, the waitress at Nasty's, "You would think a 6.1 magnitude earthquake here at New Simon would warrant violation of the University's edict of 'No Public Earthquake Predictions,' wouldn't you?"

She stared at him like he was a fool.

He blushed from the stare and declared, "It'll shake pretty strong here this afternoon, you know."

Breaking her frozen pose, she said, "Well, I sure don't plan to be here."

Now, stirring his coffee, Chris glanced at Tina bussing one of the tables and thought, at least I told her. He shook his head to rub out his worries and turned his attention to the notes for the paper he must write to gain full recognition of his successful prediction of this next earthquake on the New Madrid Fault.

Buddy Joe looked to the port side where a brown snag, a tree limb dragging on the river bottom, bobbed in the water, well away from his tow. It reminded him of his raft. His mind drifted back to the time he and his buddies horsed around along this part of the river. They played like they were Huck Finn on their raft of flotsam gathered from the riverbank for a clandestine float trip around the New Madrid loop and down past Caruthersville to Boothspoint Landing.

He still shuddered whenever he recalled how that old black fisherman, Virgil, had saved them from drowning in the wash from the big towboat. How close it had been. Then, how the old man had joked and offered them all a plug of his chewing tobacco and when they got sick, yelled at them for being so dumb. Life had been good on the river.

Jeff Collins, his new first mate, a 30-year-old who had just transferred up from Baton Rouge, stomped up the stairs. "Cable check's done, Cap'n. All barges holding tight."

"Thanks, Jeff. We're coming round Island 14 Bend. Why don't you take the helm while I visit the head." Buddy Joe stooped over so he could stand and shuffled around Jeff to the steps down to the galley. "My mind's drifting and my eyes are getting tired."

"Right, sir." Jeff noted his assumption of command in the log and pulled the stool vacated by his Captain closer to the control console so his arms would reach.

He set the book Buddy Joe had given him, *The Earthquake America Forgot*, beside the radio and looked across the farmland just visible over the trees from the wheel house window high above the deck. "You know, Cap'n, I've read most of this book you recommended on the

New Madrid fault, but it all looks so flat around here, like there ain't nothing to break."

Buddy Joe put one foot down to the second step and twisted his frame to crack his back. He said, "Well, I remember several times when I was a kid feeling the earth shake in these parts and how it worried my folks. And you know back in 1811 there wasn't much here but forests and swamp. Now there's civilization all over."

He rubbed his shoulder. "Anything man has built can be destroyed. There are buildings and grain elevators, like down there in Caruthersville, to tear up, and bridges like on I-255 downriver at the Boothspoint Landing to fall. And Associated Electric's big smokestack up at the Marston power station. Do you think it'll stay standing if the earth shakes? Even the river can break. The levees would most likely fail. Didn't you read how this wasn't even the river channel back then, and Reelfoot Lake didn't really amount to much until the earthquakes happened?"

"Maybe you're right, but I just can't believe an earthquake can be that dangerous. Nothing like a tornado or a flood."

Buddy Joe stepped down another step so he could stand fully erect. "Lots of things don't seem dangerous until you experience them, Jeff." He recalled feeling his makeshift raft pitch up and over in the wash from the towboat.

Virgil pushed his scarred wooden boat off the muddy shore next to the dirt road under the I-255 causeway. He rowed toward the big concrete pier at the base of the twin 10-foot diameter columns that supported the south end of the Boothspoint Bridge. The bridge connected Caruthersville, Missouri, to Dyersburg, Tennessee. His old fishing rod with level-wind reel lay across the side of the boat. A plastic bag of chicken guts rested in the bucket at his feet next to his tackle box.

The grizzled old man remembered growing up along the river above Memphis. His great-grandfather had been a slave in Louisiana, his grandfather a sharecropper along the Yazoo River in Mississippi. His father had told stories of the floods of 1927 and months of living surrounded by water on the levee below Greenville, of the rains, the mud, and the forced work gangs.

When they finally had the chance, his pappy and ma had sneaked onto a boat and rode it to Memphis where Virgil was born, then on to Boothspoint Landing to work on the river and in the fields. He had lived along the river all his life, watching its flood plains and banks change from wild and natural swamp to mechanically groomed fields and crafted levees.

Virgil had grown to hate the backbreaking farm work, but his passion for fishing remained the same.

For the past seven days the river had remained above flood stage. Today it flowed to within two feet of the top of the pier; he would not have to climb the rebar ladder on the side like usual. His boat bumped against the concrete pier and he tied the oarlock to an exposed piece of rebar. He hoisted his bucket, tackle box, and a folding aluminum chair onto the concrete surface.

The flat top of the pier between the columns offered access to Virgil's favorite fishing hole. From there he could cast his 100-pound test nylon line into the deep hole in the main channel, the place where the biggest and meanest catfish in the world lurked.

Picking up his rod and reel, he clambered onto the pier talking to himself. "Maybe this time, Grandpa catfish, maybe this time I'll catch you." Virgil's hand still hurt from the line cuts created when he had fought the huge catfish three nights ago.

He recalled how darkness had shrouded the river that evening, making it almost too dark to see the rope connected to his boat. After pulling the boat closer and stepping in to return to shore, he had shoved the double-aught treble hook through a chicken leg. For one final time he threw the bait with a four-ounce lead sinker far out into the river from the boat, letting the line run out from the reel as the rigging searched for the bottom.

As he unfolded the chair and placed it in the shade on the pier, he mumbled to himself, "I shore thought I'd lost another setup." Thirty seconds had passed before he felt the sinker bounce on the bottom then stop dead, like it was caught on a snag.

He had reeled the line in, drawing it tighter and tighter until he determined he would have to break it off. The nylon string sang in the river current. He had grabbed the line in his left hand and looped it three times around his right palm. He hadn't wanted to break the pole, just the line. Leaning back in the boat and pulling hard on the line, he felt it let go a bit. He gave one more tug, pulling his arm past his chest. Virgil thought the line had come free but the fish was only playing games with him. Virgil spat into the water and exclaimed, "That damned fish just took off and when he did, he almost yanked me out of the boat."

The knotted line had dug into his hand so tight that he couldn't let go, and it cut into his flesh. "Dumb, I was just dumb. That catfish could've killed me." Another jerk had pulled his forearm over the edge of the boat. Reaching into the bottom of the boat with his left hand, Virgil had groped for his bait knife, and when he found it, he lashed out, cutting the line just as another great surge pulled his armpit across the oarlock.

Virgil remembered falling back to the bottom of the boat, screaming at the pain in his hand. The line had cut deep into the flesh of his palm; it had hurt like hell. His eyes streamed tears as he sat up and unwound the bloody line from around his hand. "Damn you, Grandpa, I'm gonna get you some day," he yelled across the water.

Virgil unfolded the aluminum chair, sat down, and cut off a one-foot length of chicken guts to bait his hook, a sly grin on his face. "Today's the day, Grandpa catfish. I feel it in the air. Today's the day you'll pay for what you done."

In the galley Buddy Joe poured steaming coffee into his cup, its bottom already white with sugar. As he dropped a single ice cube into the coffee his mind's eye formed an image from the swirling cloud of steam lifting from the cup, an image of Ellen. Why? Why did she have to pack up and leave? Why did she hold it against him what he had told Jimmy? "Nothing can go wrong. Don't worry," he had told his son. He closed his eyes and shook his head back and forth, trying to clear the thought.

Stomping back up the stairs to the wheel house, he stirred the cup of lukewarm coffee in his hand with a plastic spoon. "How's it going?" he asked Jeff.

"No change. Nothing notable in the radio traffic either." Jeff made a miniscule correction to the rudder control. "Cap'n, I was wondering. When we motored north past Bessie this morning, I could see on the radar there was another tow less than a mile west of us across Bessie's

Neck, where the river runs south but it's 20 miles downstream from where we were. There's only a half-mile neck of land that separates the river from itself. Is that a place where the river levee could break and cut a new channel?"

Buddy Joe took a large sip of coffee, satisfied as the sweet lukewarm fluid washed over his tongue. "You're right, and it is a big concern at the Army Corps, though they don't talk about it much. There's a 10-foot drop in river elevation across that neck of land. If the river ever cuts through there, it'll make one hell of a rapids and stop most barge traffic until they can build a lock."

He felt a wave of resignation. It would probably take something that bad, he thought, to make Ellen come back.

Downriver at Caruthersville aboard the excursion boat *Bella Queen*, Captain Barney Ruggs leaned out the pilothouse door and over the railing to view his boat's progress as it departed the casino dock. Inside his first mate, Ralph Robinson, stood inside at the wheel. "Okay, steer the stern a little more out into the current." He watched as the stern line slackened. "Cast off the stern line," he yelled to the deckhand.

Barney could see a few souls lining the rails below. "There aren't many of our 120 passengers watching our departure this morning. They must be eating or sleeping. Everyone seemed to have a good time last night. Hope there's no one left on shore like last week." He laughed.

The young deckhand slipped the slackened rope off the dock cleat and threw it onto the deck of the boat. Then he vaulted over the top of the railing and onto the deck. "Dammit, Ralph, talk to that boy of yours. Ricky thinks he's Tarzan or something. If he keeps taking chances like he just did, he'll lose a leg or worse. I agreed to let him work as a deckhand this summer, but if he doesn't shape up I'll kick him off the boat when we get to Memphis."

"I hear you, Captain. I'll talk to him. He just doesn't understand how dangerous river work can be."

"One more chance, Ralph. That's all he gets." Barney checked that the 110-foot long boat had fully cleared the dock. "Okay, start moving the *Bella Queen* out into the current. Take it easy and slow. With the river running this high, there could be brush or old snags hanging near the bank. Once we're into the current we can move right along downriver."

Buddy Joe scanned the Hathaway mooring as it slipped by on the port side. No barges tied up for loading, so he would have the rest of the weekend free like he planned unless the company found some other short tow job. Maybe he could watch the NASCAR race on Sunday. Maybe …

The image of the black and gold racecar catapulting off the pit wall flashed through his mind: the burst of flame, the explosion, the carnage, the shock of seeing the missile tear through the neighboring pit crew. He remembered sitting in stunned silence with Ellen before the TV, waiting for word, wondering what had happened to Jimmy, the man on the jack in that neighboring pit. Their son.

The radio chattered from upriver–two other boats talking. He shook his head, struggling to clear his mind of those memories.

Jeff turned on the stool. "Cap'n, I called on my cell phone to a girl I know in Memphis. Carla Nelson's her name. She told me her brother made some kind of prediction about an earthquake that's supposed to happen this weekend. Do you believe that?"

Buddy Joe shook his head again, thankful for the interruption, and then looked askance at Jeff. "Prediction? Nobody's figured out how to predict earthquakes. Sounds like someone's making another wild-ass guess like they did back in 1990."

"But what would an earthquake do to the river if it happened? Would we be in danger here on the water?"

Buddy Joe considered the question. "We're probably in the safest place in these parts, at least in the beginning. The Army Corps book says the earth's vibrations don't get into the water, but the earth shaking can create big waves and chop. There could be some really big surges near any riverbanks that caved into the river, but I suppose our biggest danger would be from having the levees crumble all around and creating crevasses that would suck us out of the river."

"Why would the levees crumble?"

Buddy Joe looked at Jeff, such a reminder of his son. "Didn't you ever go down on a creek and jump up and down to get the sand to turn into gelatin and quicksand? If an earthquake shakes a levee hard enough, the dirt it's made of turns to mush or even quicksand, just like if you were jumping on it. And if the river's pushing against the levee, like now, it'll punch on through, and you got a crevasse."

Jeff studied a white farmhouse and barn half a mile beyond the levee. "So you don't think we'll have the earthquake this guy's predicting?"

Buddy Joe stared down the river, his mind distracted by yet more memories as he replied, "No, Jimmy, there won't be anything go wrong this weekend. Don't worry about it."

— 2 —

THE APPPOINTED TIME

Spencer Travis drove his jeep onto the levee road and closed and locked the steel gate north of Barfield, Arkansas. He settled the old straw hat on his head for his first day of levee patrol, climbed back into the jeep, and buckled in.

Shifting into first gear and letting out the clutch pedal, Spencer felt the gearbox jerk the jeep into motion as he headed down the dusty ruts to begin his inspection. He always had trouble with manual transmissions.

The river ran six feet above flood stage in this section, within seven feet of cresting the earthen dike. At that level the river water stood eight feet higher than the farmlands just to the west.

Spencer's trainer on Thursday had emphasized, "Just watch for water boils along the landward base of the levee." She pointed to a faded photograph. "Wherever there's seepage working its way through the dikes it looks like that. That's a sure sign of a weak spot that needs attention."

Spencer's whistled rendition of "Zippa-Dee-Doo-Da" competed with the sputtering exhaust of the engine as he guided the jeep along the dirt ruts atop the levee. Happy to be back at work, he bounced on the cracked vinyl seat, held in place by the new seat belt and shoulder harness. He looked over the three-foot high Johnson grass waving on the sides of the levee, toward the waters of the river to his right and the newly planted cotton fields to his left.

A rabbit darted from the grass and bounded across the ruts right in front of the jeep. Spencer slammed on the brakes, killing the engine, then laughed. "You got to be more careful up here on this levee, Mr. Rabbit, what with me driving."

Near Ditch 14A the river had touched the base of the levee for five weeks, saturating its foundations with water. In the last week it had risen closer to the top. Spencer could see that water in the bottom of the ditch away from the river seemed to be flowing, flowing toward the cotton fields.

Stopping above the ditch, he scanned the area along the base of the levee with his binoculars and spoke to the old jeep in amazement, "Well, I'll be damned. There's one of them boils. It looks just like a little artesian water spring on the side of the dike." He compared it with the faded photo. "Right, that's what it is. First day out and I've already earned my keep."

Spencer picked up the two-way radio and called the Osceola dispatcher. "Hello, dispatch, this is Spencer Travis. I'm patrolling the Barfield levee. I'm at Ditch 14A and a real live boil's coming out the bank below me." He continued with a more detailed description of what he saw.

The radio emitted static until the voice of the dispatcher answered, "I copy that. You've got an active boil at Ditch 14A above Barfield. I'll dispatch an evaluation crew within the hour to take a look."

"Great. Now what do I do? Keep going north on the levee or wait here?"

"You move on up north, and keep a real close lookout. Where you find one of those things, you most often find another."

Loretta stood in the little park on the levee just outside the Ward Street floodwall in Caruthersville, Missouri. She stared at the towboat and barges a mile and a half upriver coming along the main channel. Boats on the river had always fascinated her since listening to her dad tell of his work on river barges. Turning, she watched the *Bella Queen* excursion boat casting off from the casino dock. Interesting, she thought, if it pulls out too soon the two boats are going to have a problem navigating around each other.

She had driven that morning from the bungalow in Hayti, one that JQ McCrombie, the contractor from Memphis, kept for her. Earlier in the week she had told her sister over the phone, "Fran, JQ's real nice to me and treats me good. He's told me how he just got another big deal fixing up the highway bridges so they won't fall. And he's got a building with his name on it down in Memphis."

Fran had laughed. "Loretta, that dude's just stiffing you. He's a bastard and won't give you anything but money. No way you'll get his name. He knows what you are."

"I know, but I like to dream." She sighed. "Anyway it's nice to go into Caruthersville. I can always get better prices in the larger stores and besides, I can prospect for other business." Loretta had bowed her head. "I just wish JQ would give me more money so I didn't have to go around free-lancing. That way I wouldn't feel so much like a whore."

Her sister, also a working girl, had replied, "Yeah, tell me about it. Just don't forget who you are, babe."

Gazing back over the river Loretta saw how the water ran high, lapping at the edge of the grass next to the old cottonwood tree at the foot of the park, but that was usual in May. The paper said the river would be eight feet above flood stage soon, higher than most of the streets behind the floodwall, but the city still had seven feet of levee for protection. Plus they could sandbag the floodwall enough to ensure the safety of the town.

It felt good to rest her hip against the steel post and let her yellow dress soak up the sun. The temperature must have climbed three degrees in the last five minutes, and the humidity made her feel sticky, but she didn't have to hurry. She'd stay and watch how the boats got around each other.

Besides, one of the men at the grain elevator might be watching. She bent her right knee to assume a more suggestive pose, turned her leg out a bit, and squared her shoulders to emphasize her breasts.

Jeff steered the tow a degree to port, anticipating the next buoy. "Cap'n, take a look. There's a riverboat casting off at Caruthersville."

Buddy Joe looked up from the chart to study the activity a mile downriver. "Uh-Oh, Jeff, we've got a problem. That's the *Bella Queen* and you're right. She's just now leaving the Caruthersville casino barge. Unless Captain Ruggs gets a move on real quick, I'll have to pass that tourist boat at the same time we go under the I-255 Bridge."

Buddy Joe reached for the radio microphone. "But if he'll just stay where he is, at this speed we'll close in seven minutes." Flicking the switch to the local frequency and pressing the button he spoke. "Captain Ruggs, this is Captain Simpson of the *Lady Bird Jamison*. We are cruising downriver a mile and a half above your current position. Please hold the *Queen* in place until the *Lady Bird* is clear."

On the *Bella Queen* Barney listened to Buddy Joe's voice from the speakers on the wall of the wheel house ordering him to stay in place so the towboat could pass. He had started to speak several times, unsure of just how to put what he had to say.

"Now, ah, Buddy Joe, ah, hold on there." He waited to be sure Buddy Joe would be quiet. "The folks on this boat would be very upset if I fell in behind a towboat. They'd wonder why I didn't get ahead of it. Buddy Joe, it's a matter of principle. I'm pulling into the main channel now."

The voice on the radio came back. "Well, I don't plan to slow down, so you better get your little tourist boat headed downstream pretty fast. I'm moving at 13 knots with this current, but I'll stay half a mile behind you if you'll get a move on."

"I'll do just that, Buddy Joe." Barney's red face reflected his response to the pressure. "Just keep your temper down and I'll keep out of your way. Besides, it looks like it'll be a nice cruise down the river today, don't it? And maybe we can talk along the way." Barney and Buddy Joe often chatted when they met each other going up and down the river. Today would be one of the rare times when they would travel in the same direction and remain in close contact for the entire day.

Buddy Joe sounded gruff but laughed. "If you say so, you're the one who does all the talking. Just get a move on." Barney smiled and relaxed when he heard Buddy Joe chuckle.

Jud stepped back from the loot he had laid on the dirty blanket on the side of the slough above Osceola. He scratched his elbow and watched warily as the scruffy fat man, his hairy stomach falling from beneath his khaki shirt over his belt, poked through the assortment of fishing reels and boating equipment.

Jake said, "This all you got? I'd a thought that in a couple a weeks you could pick up lots more than this. There ain't more than 50 dollars worth of junk here." He spat tobacco juice into the nearby poison ivy.

"Jud, you're spending too much of your time popping meth, drinking beer, and screwing your 12-year-old sister."

"Only 50 dollars? You gotta be crazy, Jake." The thin, pimply-faced 17-year-old reached down to scratch the "speed" bumps on his knee through his torn jeans. His tennis shoe rested in the mud next to a beat-up johnboat pulled partway onto the bank.

"I cain't just breeze into any marina and go through every boat. I got to keep a low profile and search the empty boats after it gets dark. It ain't easy finding stuff like this along the river." He sniffled and wiped the back of his dirty hand across his nose. "I figure there's at least 70 dollars worth of good stuff there. You'll be able to sell it for a two or three hunnerd at your second-hand store in Osceola."

"Fifty-five."

"Sixty-five."

"Fifty-five, take it or leave it." Jake pulled out his billfold and peeled off two twenties and three fives. "Do you want some beer? I fenced a few cases from Billy Jack last night. I kin give it to you for half price, two-fifty a case."

"Sure, I'll take a couple a cases."

Jake put a five back in his billfold and handed the remainder to Jud. He spat again into the poison ivy.

Jud walked to the back of Jake's rusty pickup and lifted out two cases of beer. Walking over to place them on the floor of his boat, he looked up to see Jake waiting with hands on his hips. "Whadaya want, a free beer?"

Tina, the waitress, stopped in front of Chris holding the coffepot. "Warm-up?"

"Sure, why not?" He stared into his cup, watching the cream and coffee mix in a swirl. The girl moved down the counter to serve a couple of truckers talking baseball. Chris looked up to see the cook place an order on the serving shelf under the red lights. The second hand of the clock on the wall behind the old man moved past 12. The big hand pointed to 34.

Nine seconds later, at 9:34:09, the earthquake that Chris Nelson's program had predicted became fact. Eleven miles beneath New Simon the rock matrix ruptured, starting a fracture that over the next 23.5

seconds would grow along the southern extension of the New Madrid Fault, reaching southwest to Lepanto, Arkansas, and northeast to Ripley, Tennessee.

In the next four minutes the Mississippi and Ohio Rivers would break from the quaking. The lands from St. Louis and Cincinnati to New Orleans would be shaken to destruction.

In less than 30 seconds, Chris, rushing back to his seismic station on the other side of the interstate, would lie in his crushed truck beneath a fallen overpass, a casualty of his own earthquake.

Earlier in the morning aboard the *Bella Queen,* a young man rolled over half asleep and encountered resistance. When he noticed he lay on a small bed next to a naked lady, it took a moment before his drowsy mind remembered that he and Lynn occupied a riverboat cabin on the Mississippi River at Caruthersville. Today their boat would depart downriver toward Memphis for an evening's entertainment at the Jazz Festival, then return on Sunday.

Ron Cannon smiled as he recalled the evening before. What a joy, at least it had been once he awakened from his stupor. Things had been touch and go after he departed the open house party in Memphis yesterday afternoon. He had a couple too many drinks while listening to that fellow Chris Nelson talk about the chances for a big earthquake on the New Madrid, and he had arrived late at the airport to pick up Lynn. But later, before falling asleep, he declared his love to her, so she had been forgiving when he awoke.

Lynn Browne, an Air Force Captain stationed at Scott Air Force Base east of St. Louis, and Ron, a retired Air Force Major and former C-17 pilot now with Federal Express at the Memphis International Airport, had worked together for the past two months. Their assignment had been to prepare a joint contingency report detailing how the Air Force Logistics Command and Federal Express should handle a catastrophic event, such as a terrorist attack or an earthquake on the nearby New Madrid Fault.

Both recent divorcees on the bounce, they had let their close working relationship develop into something more. Having finished an assignment the previous week, they arranged to celebrate the report's completion and test their personal relationship with a weekend outing on the river.

Pushing on her shoulder in the early morning light, he had whispered, "Hey, sleepy head, wake up. We've got a whole new day ahead of us." Lynn roused, turned her head toward his face and put her arms around his neck, surprising him with a kiss of fiery rapture.

An hour later after they slipped out of bed, Ron stepped into shorts and a tee shirt and stuffed his toes around a pair of thongs. "Let's get out on deck and go get some breakfast. I could eat a horse."

He glanced out the porthole on the starboard side of the boat. He saw the riverboat pulling away from the Caruthersville dock next to the casino, en route to Memphis. Checking his watch he exclaimed, "My God, woman, it's after 9:30. They close the galley at 10:00. We'll starve."

Scanning the scene through the porthole glass he watched as the river changed. Its surface began to seethe with a frothy spray.

Barney walked to the other side of the pilothouse high atop the *Bella Queen* and looked back toward the approaching *Lady Bird Jamison* and her tow. "We're clear of the slack water, Ralph. Go to full throttle." His mate eased the throttle forward and the twin 240-horsepower diesel engines increased rotation speed on their respective screws to accelerate the boat to cruising speed.

At 9:34:13 the leading edge of the P-waves from the fracture coursed through Caruthersville. It took Barney several seconds to become aware of the transformation in the river surface. One moment the river appeared smooth and light chocolate brown with swirls rising to show bumps and hummocks in the river bottom 40 feet below. Then, as if a sudden windstorm had struck, the surface of the water rippled and whipped into foam.

P-waves from the fracture moved up from the river bottom and spread into the water as low frequency, high-volume sound waves. They made the surface dance. Droplets jumped out of the water only to fall back like raindrops. Tiny wavelets appeared, forming steep-sided cups on the surface. Chop that moved in all directions appeared, as if by magic, on the water.

Grabbing the microphone, Barney radioed, "Buddy Joe, what's happening to the river? The water down here just took on a life of its own."

Five seconds later S-waves coursed across the river bottom. Unable to transmit seismic energy into water, a material with no shear-strength, the S-waves could only pummel the mud and sand of the riverbanks and bottom and then reflect back down into the earth.

Reflecting S-waves physically moved the interface between the river water and its banks and bottom back and forth and then up and down, like some giant shaking a shallow pan of water. A violent mixture of two- and four-foot waves raced from the banks across the surface of the river. At the same time the river bottom churned up and down, creating a plethora of violent swirls, eddies, and currents.

Buddy Joe's voice yelled over the radio. "Barney, the river has gone mad. I see a giant chop growing, and we're picking up some really humongous waves. They're coming from all over, big time."

A witch's brew now surrounded the riverboat, barges and towboat. Barney heard Buddy Joe suck in a deep breath. "We must be having an earthquake. That's the only thing that could be shaking up the riverbanks and bottom this bad."

Barney looked around and thumbed the mike to report. "You're right. I see trees and power poles in Caruthersville shaking around, but there ain't no wind."

Buddy Joe's voice continued, "… my barges are doing a hellish amount of twisting. Barney, I'm reversing props and will try to hold in this position. I'm praying my tow holds together. This pitching is putting a terrible strain on the wires and ropes. Every barge wants to go its own way."

"Buddy Joe, I know what you mean. The *Queen* is already tossing about in these waves, worse than a hurricane. Waves from off the bank are slamming us big-time. We're only 30 yards from shore and it's beating the hell out of us."

Barney looked back toward the casino. "Damn. The dock has broken apart and the levee is settling. Now the casino boat is capsizing."

"Barney, get away from shore as soon as you can and head for deeper water. Get away and hold there. I don't know what will happen next, but I don't think things will get better any time soon." Buddy Joe's voice sounded as strained as Barney felt.

"Roger that. Just hope I can get there without dunking some passengers or crew." He turned to his man in the wheel house. "Ralph, we have a General Alarm condition. Steer this ship to the middle of the river at flank speed. Bring it around to head upstream and hold. I don't want to move up or down the river at this time."

Barney hit the General Alarm button with the heel of his hand. The siren sounded throughout the ship as he picked up the intercom microphone and shouted. "Ladies and gentlemen, your attention please. This is not a drill. This is a General Alarm. Repeat, this is not a drill. All passengers immediately put on your life jackets and go to your lifeboat stations to await further orders. The land around us is having an earthquake and we don't know what its full effects will be to us on the river.

"I repeat. This is not a drill. This is a General Alarm. Put on your life jackets and go to your lifeboat stations immediately. Directions to your stations are on the inside of your stateroom door. Move quickly. Now." The warp of the siren echoed back to the ship from the riverbank.

Even as the Osceola dispatcher was telling Spencer, "You move on up north along the levee and keep a real close lookout. Where you find one water boil, you most often find another," thunder rumbled from the ground. The surface of the river to his right turned dull, no longer reflecting the sky and trees from the banks and hills across the water. The levee began to sway.

Spencer asked, "Dispatch, this is Spencer again. The levee's shaking, real hard. Does a boil make it do that?" Static obscured any answer.

The fracture in the crust passed nine miles west of Barfield. The leading edge of the S-wave reached the levee under Spencer eight seconds after the start of the rift.

"Osceola. Can you hear me?" Spencer's voice warbled as he yelled into the microphone.

The Barfield levee had been constructed in the same manner as all the levees along the banks of the Mississippi River from above Cairo, Illinois, to New Orleans. Barfield's levee just happened to be the closest to the epicenter of the earthquake, the first to be stricken. Composed of the same Mississippi sand and mud that make up the farmlands of the former flood plain of the river, the Barfield levee loosened and crumbled from the shaking and sagged to its lowest stable level.

"Osceola. The levee's starting to sink. Osceola." Spencer shoved the accelerator to the floor and let out the clutch of the jeep, gathering speed along the levee road, bouncing in and out of the ruts.

Where water had saturated the soils of the levee, as evidenced by water boils that search out such weaknesses, the loosening and crumbling spread even faster. The levee became the consistency of quicksand. It did not sag. It flowed. Within 13 seconds the water pressure of the river punched through the earthen dam and gushed toward the land to the west.

Spencer drove like he had never driven before, screaming into the microphone. A crevasse opened behind him, then 100 feet in front of him. His jeep dropped into the quagmire to be swept along by an eight-foot deep wall of water from the river. The jeep tilted, then rolled. Water poured through the open passenger compartment.

Too late, Spencer grappled against the rushing waters to remove the seat belt that trapped him inside the tumbling vehicle.

— 3 —
The Fracture Grows

Steve Pauli drove his old pickup truck north on Tennessee Highway 22 from Tiptonville toward the old landing town of Bessie on the Mississippi River. His daughter Juliana bounced on the slick truck seat beside him as he maneuvered around the chuckholes in the old pavement.

"There, isn't that a pretty honeysuckle vine?" Steve pointed out the scattering of delicate white flowers near an old farmhouse, the last habitation he could see as they approached the base of the New Madrid loop.

Glancing out the corner of his eye he could see that Juliana wanted him to understand that she did not think this was a fun trip. She turned her head too late in the direction he pointed and said, "Oh, guess I missed it."

Steve felt the alienation. He wondered how he would ever convince his daughter to enjoy being with him when his visits were limited to only two weekends a month. He spent most of his time earning a living and paying child support. His second job as a free-lance photographer cut into his spare time, leaving little time left over for fun.

Now he drove along the old road on an assignment to shoot background photos on the half-mile wide isthmus that separated the two bends of the Mississippi River, Winchester Towhead to the east and Point Pleasant Chute to the west.

"Hey, see that ground squirrel crossing the road up ahead? He's really scooting along."

"What squirrel?" Juliana stared at her lap and turned her hand to inspect the purple nails she had applied that morning.

The paved road veered left, away from the levee, but Steve steered the pickup onto the dirt track to the right. "We'll pop up over to the other side of the levee to a turnaround beside the river where I've done some great photo shoots."

He drove 100 feet along the top, then turned right and followed the ruts that led down to the river. As the dirt road dropped into the trees and rounded the turn, he jammed on the brakes. "Holy cow. The turnaround's covered with water. Nobody told me the river was running this full."

East of Bessie's Neck the river flowed north, its swollen surface at an elevation of 298 feet above sea level. Behind Steve, the top of the levee holding the river in its channel stood at an elevation of 305 feet. A seven-foot high wall of sand and mud protected Bessie's Neck from being cut by the flowing river.

"Guess I'll have to back out this time, but we can still get some good pictures," Steve said as he set the brake. "Come on, let's get out and walk around."

"Do I have to?"

For God's sake, Steve thought, now Juliana is pouting and slumping her shoulders, just like her mother does, to show how weak and tired she feels. This is too much.

"Yes, you have to. Come on. I only get to visit with you two weekend days a month and I don't intend to waste any of that time with you sitting in this truck while I take pictures. Now move your little butt out that door." He grinned to show the big bluff behind his gruff voice.

Juliana could not help but grin back. "Oh, Daddy, but really, this trip is just 'soooo' boring."

Loretta watched the towboat and barges coming along the main channel. Her hand reached for the post set in the levee in an instinctive response to the P-wave vibrations. With her thoughts focused on the river, she ignored the growling sound of a distant, roaring train and the tickle in her sandaled feet from buzzing in the asphalt walkway.

The levee jerked away from the river, then back, snapping her attention to where she stood. It jerked again and continued its erratic behavior. The pace quickened and the motion became more and more chaotic. Buildings along Third Street made clanking and breaking sounds as they cracked apart from the shaking. The 90-foot high concrete grain elevators next to the river rocked back and forth until one burst, starting a shower of ochre-green soybeans mixed with cattle-sized chunks of gray rock.

Ten miles beneath the surface, the leading edge of the oval-shaped fracture passed under the river two miles southeast of Caruthersville. The earthquake's magnitude had reached 7.3 and continued to grow.

Loretta stared as large waves washed up from the river and splashed onto the shore. She spied cracks developing below the park in the levee revetments, the concrete coverings that shielded the dirt from erosion. The slabs began to slip toward the water. Looking farther downstream, she saw the old pump house and other levee structures slip and begin a lemming-like march into the river. The very ground on which she stood appeared to be sinking as the water swirled up the asphalt walkway toward her feet.

The water table in Caruthersville averaged two feet beneath the ground's surface. After only seven seconds of shaking, the granules of sand and silt between the layers of clay became suspended in the ground fluid, a process called liquefaction. Down to a depth of several hundred feet, the whole area under the town mutated into a thick flowing ooze, like quicksand, into which heavier objects could sink.

Pressure in the quicksand from the earth's shaking and the weight of the clay layers on top forced the thin soupy mixture up through cracks to the surface. In some places it erupted like a geyser, spewing dirty water, sand, and the waste of centuries out over the land. The humid stink of rotten eggs and sulphur filled the air.

Before the earthquake began, the highest point in town rested at an elevation of 279 feet. During the shaking, its elevation dropped six feet. Building foundations sank as much as 12 feet more into the mud.

Trickles of muddy water began to flow over the top of the levee, cutting channels in the dirt. Downriver, south of town where the farmlands were lower, the trickles became torrents.

Loretta clung to the post for support. In fascination she observed the actions of the land around her for several seconds. Then, realizing the awful truth, she turned and ran back into town along the drunken pavement, without considering where she might go. A blinding flash and explosion behind her came as grain dust in one of the concrete elevators ignited from electrostatic sparking generated by the shaking.

By now she had lost her sandals and ran barefooted, flailing her arms to keep her balance on the bouncing road, dodging falling bricks, light poles, and writhing wires, screaming to everyone she saw, "The river's coming. The river's coming. It's coming over the levee." She prayed she could outrun the advancing waters and falling debris.

From their anchorage above Fulton, Tennessee, Sylvie Green scanned the trails along the edge of the bluffs, six miles below Osceola. "Charlie, up there in the trees. Look, I see a couple of girls riding bikes on the trail. Do you see them?" With binoculars pressed against her eyes, Sylvie sat in a canvas chair on the fantail deck of their 39-foot motor yacht, the *Amanda Blair.*

Charlie rested his elbow on the railing, balancing the fishing rod across his fingers. He squinted and looked up as the early rays of sunshine highlighted the trees in Fort Pillow State Park 300 feet above the Mississippi River atop Chickasaw Bluff Number One. "Kind of hard for me to see them without the binoculars. I'll take your word for it."

Sylvie watched him lift the fishing rod to feel for a fish and then drop it back down to rest on the side of the boat. He didn't really seem to be looking very hard.

The evening before, they had left the dock at Memphis's Wolf River Marina in the boat they had purchased in New Orleans. Sylvie had told her husband of 48 years, "Charlie, this is fine. I really do like fishing trips. I didn't want to go cruise the Caribbean anyway. You made our fortune and now we can spend it however you want, even bottom fishing in the Mississippi River if that's what you must do." She smiled and added, "Just so long as we do it in grand style from this air-conditioned cabin cruiser."

The girls reappeared in a clearing on the high riverbank, south of the preserve commemorating the bluff's role in the Civil War. "They must be the same age as Amanda and Blair. Our granddaughters will be 14 and 15 this year, you know." Sylvie dropped her head to her hands and tears streamed from her eyes. "Oh my, I just can't keep from crying."

Sylvie heard her husband speak. "Now, Sylvie, you know it doesn't do you any good to think about the girls. Their mother has decided they're lost to us. We won't see them again, unless they want to find us someday when they're grown." She looked up to see through her tears a concerned frown covering his rough face. "Just don't get yourself all upset about it. The doctor says it's not good for your heart."

Sylvie sighed and again lifted the binoculars to watch the two girls as they appeared from time to time along the crest of the bluffs. She waved to them, hoping they could see her. "Whose granddaughters are they? Do you think they ever visit their grandparents?"

Charlie shook his head.

"Maybe someday I'll have my grandbabies back," she told Charlie. "Maybe I'll be able to hold them close to my breast and hug them like I so want to do." Sylvie dropped her head again and let the tears stream from her eyes in silence. Charlie gave her everything she had asked for, but he had failed to give her what she wanted the most. Sylvie felt so sad, so alone.

Jake finished the last of his beer and threw the bottle into the slough. He burped then asked, "You seen that big cabin cruiser downstream that's been anchoring overnight off Fulton the last couple of days?"

Jud looked up as he wiped his mouth with the back of his hand and belched in reply. "Yeah, what about it?"

Jake gathered the blanket of stolen gear into a bundle. He cocked his head with a sly glance out the corner of his eye. "There's an old couple that owns it. The old man made his fortune in soybeans and likes to fish off the bluffs at Fort Pillow. They must have lots of cash aboard, because I hear when he docked at Osceola he paid in cash from a big bankroll. You could find a real treasure there."

"What? Are you saying I should rob them?"

"Why not?"

"I ain't never robbed anyone direct, Jake, and 'sides, what if they seen me?"

"The river kin swallow up lots of troubles if you want, and I might find a buyer for that yacht if you could hide it up here in the slough." He watched Jud in expectation.

"Well, I don't know. I ain't never done anything like that." Jud leaned over to slide his dirty boat back through the mud into the water.

Jake spied a tree frog hopping next to the poison ivy. Scratching his belly he spat a stream of tobacco juice across its back and chuckled, watching it squirm and spasm through the mud. "Well, think about it. You ain't getting rich on fishing gear you know, and with them old folks gone there's a bunch of people who'd really like to have that boat."

Placing the blanket bundle in the bed of his pickup and throwing a tarp over it, he opened the cab door to clamber inside. "See you around in a couple of weeks unless you call me sooner. Same place, right here."

Jud said, "Okay, I'll think about it. And thanks for the beer. I'll go share it with my sister." He smiled as he stepped into the johnboat and shoved it into the water. "Beer makes her playful."

Jake grinned at the vision of Jud and his sister doing it. He watched Jud start the outboard motor and steer back to the center of the slough, then accelerate toward the river. Strange, he thought, the wake of the boat was making the slough spring up little waves all over.

Daydreaming on the south support pier of the I-255 bridge over the Mississippi River, Virgil snapped awake when the P-waves swept across the river. Remembering the jolt he had felt from the earthquake last weekend, he jumped fast, for an old man, and grabbed the rebar ladder on the bridge column. Five and a half seconds later the bridge column began jerking about with accelerations strong enough to throw him into the water had he not held on.

The shaking worsened. "Oh, my God. I don't believe this, God. Please stop," Virgil called to the heavens. He felt the column shudder north, then spring back. It lifted, and then sank as the water of the river rose and flowed around his feet.

The chop and growing waves of the river washed his boat about downstream from the column. His lawn chair bounced and flipped over the side before it sank in the muddy waters.

Virgil looked up as the bridgeworks above him began to make cracking sounds. He stood on the south most of the three main piers buried 180 feet into the bottom of the river. A pair of concrete pillars, 10 feet in diameter and 80 feet tall, rested atop each pier.

On the Tennessee side of the river, two sections of 300-yard long cantilevered steel truss sections rested on the pillars and stretched over the deep-water channels.

Stretching northward from the trusses, ten smaller versions of the primary bridge pillars carried 100-yard long causeway sections of the four-lane divided highway over the flooded woods of Island 17. This half-mile long causeway connected the cantilever truss to the embankment on the remains of Island 16 where the freeway continued on to Caruthersville.

On the Tennessee side a suspended 50-yard steel plate above Virgil connected the truss to three spans of causeway similar to that on the Missouri side. Beyond that point, another three quarters of a mile of causeway rested on pre-stressed concrete joists held aloft by concrete pillars like many of the freeway overpasses along the interstate.

All across the bridge the roadway consisted of separate northbound and southbound segments of pavement. Segments averaged 50 yards in length with periodic joints provided for heat expansion and contraction. Roadway segments rested on the under-structure, but since they moved back and forth, the couplings connecting one to the next had been made loose on purpose. Nothing in the design of the couplings to the undercarriage contributed to structural integrity against lateral forces on the bridge.

S-waves from the earthquake, followed by the slithering Love waves and the rolling Rayleigh waves created when the S-waves reflected back into the earth at the surface, generated horrendous forces and motions in the soil in the flood plains and beneath the river around the bridge.

Accelerations at the base of all the columns exceeded that of gravity. The columns, towers, and bridge piers that were buried in the mud felt the same kinds of forces as tall buildings on dry land. Their predicament worsened because the primary columns rested in mud rather than on bedrock and the roadway tied their tops together.

The initial S-waves moved the earth back and forth parallel to the fault line in a direction perpendicular to the direction of the bridge. While the motion rocked the road surface back and forth and threw autos and trucks to and fro between the lanes, the bridge structure itself remained solid. These S-waves did little damage to the bridge, but the seismic shaking turned the dense packed mud around and under the support columns into a runny broth. The bridge began to resemble loose teeth under a flexible dental bridge set in gums that had dissolved.

When the Love waves pummeled the bridge six seconds later, their lateral movement relative to their direction of travel paralleled the direction of the bridge. The loosened support columns felt accelerations in the direction of the roadway, which stretched and compressed the distance between neighboring support columns.

The S-waves did not stop when the Love waves appeared. The two waves mixed together and formed a chaotic soup of motion in both directions. And, as if that were not enough, they combined to form Rayleigh waves with their characteristic up, forward, down, backward rotation.

The entire bridge writhed like a worm on hot pavement, screeching in agony and tearing itself apart.

Within eight seconds the first failure in the bridge structure happened one mile north of Virgil. This occurred at the point where the end of the causeway touched down on the dirt embankment of the freeway right of way in Missouri. Subsidence of the embankment stretched the spacing of the southbound lanes beyond the length of the support beams and the roadway dropped into the flood plain below. It's collapse left a 20-foot precipice. Three cars flew off the end of the road before motorists still speeding south could be warned about the danger.

After 15 seconds of shaking, the second failure in the I-255 Bridge occurred above Virgil. The Tennessee causeway leaned south just as the main truss section of the bridge accelerated north. The terrible strain at the end of the connector plate joining the truss to the causeway pulled apart the rivets holding the two together. The south end of the southbound lanes of the connector section ripped loose and fell to the ground below, leaving the north end of the section propped against the main section of the bridge. Virgil's old pickup parked on the side of the road beneath the southern end of the steel plate disappeared in a cloud of dust.

Virgil clung to the rebar ladder watching the turmoil around him. He remained speechless until the falling causeway drove his parked pickup 15 feet into the sand. Then he cried, "God, stop this and please save my soul. I'll be your greatest disciple or whatever you want, I promise, but please, Lord, just make this shaking stop." He turned his face to the heavens, as the shaking grew more intense. "I don't deserve this. You know that, Lord, I swear I don't." He screamed even louder. "I know I've sinned, but I sinned only a little, not this much. Oh God, please make it stop." Motion sickness overcame the old man, and he puked onto the heaving concrete.

The shaking slowed. "Thank you, Lord. Thank you." Virgil felt sure that God had listened and answered his prayers. But the quiet only heralded the end of the seismic waves coming directly from the fracture. Four seconds later, seismic waves reflected from the earth's mantle 60 miles below began pouring through the river channel, continuing for another 30 seconds at almost the same intensity as before.

The three bridge piers supporting the truss sections over the deep-water channel reached the deepest in the river bottom. The piers went all the way down to compacted layers of mud and sand where they sat atop concrete pilings driven even deeper into the alluvium. As the concrete piers twisted and shook and twisted and lashed back and forth, they drove the pilings deeper by another three feet.

The steel truss resting upon the support columns stretched and then compressed. Rivets sheared and popped out of their retaining holes at the most severely stressed parts of the arches, but the arches held together even as the seismic waves intensified.

Virgil watched the splash of water blow out as the southbound causeway reaching into Tennessee collapsed. With no more support from the missing connector plate, the causeway columns were pushed north back toward the bridge, and they leaned over in the soft liquid mud around their base.

As the weakened columns pushed on their neighboring columns down the line, the whole line leaned more and more, until all the columns collapsed together onto the floodplain below. Cars caught in those sections fell with the roadbed. When the road hit the flood plain below the cars bounced high into the air, and then disappeared into the enveloping dust and water.

As the causeway fell, one car continued speeding across the bridge from the north, finally reaching the place where the fallen plate above Virgil had once been connected. It shot off the end of the causeway over the roadbed, which was now half buried in mud. The car struck the fallen causeway and rolled over the side of the concrete roadbed away from Virgil's buried pickup. The car burst into flames. "Oh Lord,

I pray for their souls. Oh God, please stop." Virgil's voice grew weaker and weaker as he clung to his support and watched the world collapse around him.

But the destruction continued. A tremendous crack of popping rivets gave notice to Virgil that a fifty-yard section of southbound roadbed in the main truss section over the river had broken loose, first at its south end, then at its north end.

The concrete roadbed tore out the beams beneath it, and like an Olympic diver doing a half gainer with a twist, plunged edgewise into the river channel below, 100 feet away from Virgil. Disappearing into the depths of the river, it threw a huge splash of water towards the Tennessee shore. The 20-foot wave crossed the channel and washed up and over Virgil, almost dragging him from the rebar ladder to which he clung.

"God, oh God." Virgil sputtered at the top of his lungs. "Please don't do this to me. I promise I'll be good from now on. Please stop."

The shaking stopped. The waters calmed to a mild chop. Virgil dripped of river water and stood, clinging to the rebar, shaking like a leaf, in three inches of water atop the sunken pier. His half-filled boat waved about in the current downstream from the concrete pillar.

The shaking of the earthquake had torn at the bridge for 79 seconds. Virgil's prayer had been answered but not before the entire southbound lane of the bridge, from Missouri to Tennessee, lay in ruins beneath the water.

— 4 —

Destruction Along the Rivers

Cairo, Illinois, built on the five-mile long peninsula above the juncture of the Ohio River with the Mississippi River, sits on the flood plain next to the Ohio. It is protected from the Ohio floods by a moderate levee and by flood control dams on that river's tributaries. It is protected from the Mississippi floods by a major levee along the west side of the town.

Three miles below Cairo at the tip of the peninsula, 19-year-old Paula Kelly rested on an old rusty bench and stared across the Ohio from Fort Defiance State Park. She watched pieces of flotsam drift downstream into the eddies where the flood waters of the two mighty rivers intermingled below the old fort, "Ohio blue with Mississippi chocolate" as some people called it. Those words gave Paula solace.

She had told Freddy, "I like to think of things coming together, becoming part of the whole, different but then one." Somehow, the analogy between this place and her life with Freddy seemed appropriate. "Freddy, you're black, I'm white, and when our baby's born next week, he'll be muddy, just like the river downstream."

Paula knew her parents would never give in, never accept the fact that she loved Freddy, even though he was different, or maybe she loved him because he was different. She dared not care anymore.

"One more week," the doctor said. "But the way you're dilated the baby could come at any time if it decides to make its move, so don't go out and around. You stay close, just in case." From Fort Defiance she could drive home, three miles away across the Ohio in Kentucky, or to the hospital in Cairo, five miles from where she sat. Freddy worked at the Shell station just up the road and could come immediately if her labor started.

Paula felt like crying, but that wouldn't do any good. She told herself, I just have to find a way to work through this. Freddy is committed to me, and I can live without my parents.

They had walked down the path to the Ohio, its waters now running high against the bank because of the flood in the Mississippi River. Paula looked out at the barges moored along the sunken shore. "You know, I'd like to get on one of those barges and push off into the river, go far away from here. Far away from all the stuff where I grew up and my folks and all the people who think we're wrong. Maybe someday we can just float down the river away from all this."

Freddy hugged his girl-woman, the mother of his child-to-be. "Sure, honey, sure. Some day we'll just go away from here."

A growing cloud passed in front of the warm sun, casting a shadow across the park. Paula leaned to the side and reached around her extended belly to pick up a stick. She doodled a design in the sand at her feet, waiting, thinking about going away.

Above Caruthersville the river became a maelstrom, tossing the barges of the *Lady Bird's* tow up then letting them crash down. The land shook and jerked beneath Caruthersville and the Mississippi River with an intensity sufficient to destroy most buildings and infrastructure and with accelerations at a level approaching that of gravity.

The P-waves and S-waves often doubled their amplitude as they reflected downward while crossing the river bottom, mixing the mud and its cargo of pollution with the water of the river. Buddy Joe held his hand on the large red Collision Horn button while chop and whitecaps, uncertain of where they wanted to go, leapt in all directions from the surface of the river. The "whoop, whoop, whoop" of the horn warned the deck hands who had been inspecting the bindings of impending danger, telling them to find a place to hang on for dear life.

"Watch out for broken ropes," he yelled over the intercom. "Ride out the storm. This is an earthquake. It's trying to tear us apart. Don't get hit by a flying rope." Turning to Jeff at the helm, he said, "Reverse props, hold everything right here."

As the Love and Rayleigh waves tore across the river, they sloshed the water of the swollen river back and forth between its already full banks. The 12 barges responded to the waves by moving up and down, but they did so as individual hulks. The deafening noise of barges scraping against each other almost obscured the rifle-shot sounds of several one-inch steel ropes parting from the strain.

"Don't get caught between two barges," Buddy Joe yelled into the intercom. "The barges are rocking from an earthquake. Hang on." His grip on the window jam turned his knuckles white.

Boats and structures on or near the riverbank, like the casino barge and the grain loading booms, all failed in the first 20-second pulse of seismic energy from the northern half of the fracture. It tore through the land but then the quaking calmed for four seconds while the water continued to slosh from its own momentum.

Buddy Joe watched in awe from the bucking wheel house of the towboat as his raft of barges strained to break apart. Cracks slashed across the glass of two windowpanes in the wheel house. Three more ropes in the center of the tow parted. A five-inch diameter cleat on the port barge tied to the towboat with a steel rope bent over and was then ripped from the steel deck.

The steady pounding of seismic energy from the southern part of the fracture and the reflected seismic waves from the mantle swept through the region for the another 50 seconds, building higher and higher swells in the river. The battle waged on, each new convulsion stretching the remaining ties farther and farther.

Everett had taken his time driving from the campground on the west side of Kentucky Lake across the long earth-filled portion of the Kentucky Dam blocking the Tennessee River. He had stopped at several observation pullouts to investigate the dam and take pictures of shore birds on the riprap rock at the water's edge below.

As he left the last turnout next to the concrete core containing the floodgates and hydroelectric generators, his wife Deb commented, "This dam's not high like Hoover Dam on the Colorado, but it surely looks big and fat."

"Yup, it makes up for its height by its length and holds back a tremendous amount of water. This earthen portion we just came across is over a mile long, and the concrete section ahead is about a third of a mile. This and the Barkley Dam on the Cumberland make up the Tennessee Valley Authority's primary hydroelectric project."

After viewing the Army Corps of Engineers display in the Visitor Center and learning about the importance of the lakes to the nation's water transport system, the couple drove across the highway to the barge-lock viewing area.

Everett leaned on the shiny metal railing of the observation platform and looked down at the three-by-three rack of barges loaded with coal. The towboat settled the barges into the lock basin with gentle care.

Pointing back towards the lake he explained to his wife, "Deb, in a couple of minutes now you'll see the upper gates start to close. When they finish closing, the lock master will let the water out of this lock until it is level with the river below; then he'll open the lower gate and let the towboat push its load out." He chewed on a blade of grass he had found as they had walked from the parking lot.

"How long do you think that bunch of barges is?"

"I saw over at the Visitor Center that barges are mostly a standard size, 35 by 95 feet, so let's see, that group must be 285 feet long and 105 feet wide. There's not room for anything much bigger. You can east of here where they're constructing a new lock that will handle bigger racks."

With his digital camera Everett snapped a photo showing the length of the huge dam that held back the Tennessee River on the other side of the lock channel. "The lake waters are at the top of the flood gates, so the lake is full. They're having lots of rain this spring in eastern Tennessee. I saw it posted that the lake waters are 92 feet above the river level below the dam today. Those lock doors must be at least 120 feet deep."

"Everett, can we drive on down into the Land Between the Lakes after we've checked out these locks? I know it's only 9:30, but I'm already hungry for a sweet roll and a cup of coffee."

Everett zoomed his camera in to take another picture of the towboat. "Yup, we'll go in a couple of minutes. I want to get some pictures of the lock closing behind the towboat."

Deb sat down on the bench to wait and opened her knitting bag. Everett felt pleased she knew how to wait and turned back to adjust his camera.

Steve Pauli, squatting next to a fallen tree that stretched into the floodwaters of the river, pointed to a strand of unfurling green fern to its side. "Now what do you think? Will a close-up of that leaf make a great backdrop or not?"

Juliana stepped back for a critical look. "Well, if you use the dark bark of the tree as a background." She reached down to pluck a small flowering weed. "And add this little flower, it might be nice."

"Say, maybe you take after me after all. You have a good eye for arranging a photo."

Steve positioned the camera for the first of his five shots. The fern cluster moved, then the camera started vibrating and Steve could not get into focus for the close-up. "Dammit, hold still." He tried to talk the plant into posing.

"Daddy, what's happening? The ground is trembling and I feel thunder." Juliana stepped back, tension in her voice.

Steve raised his eyes to look around. He too felt a vibration in his knees and heard a low-pitched rumble. A surge of fear darted up his spine. A furtive menace surrounded them, something he could not see but could feel, something terrible.

"Come on, Juliana. Let's get back to the truck." He rose and reached out. Taking her hand, he hustled back up the dirt road away from the water.

The levee where Steve and Juliana were taking photos was 39 miles from the epicenter of the earthquake. The fracturing had almost ended by the time the earthquake jerked the ground beneath their feet to the north then to the south. Trees around them snapped and pounded each other. The erratic shaking grew and flung the pair to the ground. Steve crawled on hands and knees tugging Juliana up the levee road, his camera dragging in the dirt.

A slithering east-west motion accentuated the north-south jerks. The land heaved upward, throwing them into the air. The pair could find no way to hold onto the earth. For 11 seconds the land shook harder than any roller-coaster ride Steve could ever remember. Then it eased.

"Daddy." Juliana screamed. "The river's coming at us. We're going to drown."

Steve turned to see the surface of the water swirling with a chaotic chop and shooting over two feet high. The river seemed to be climbing the side of the levee. The water that had been 50 feet away now splashed within 10 feet. Even as Steve watched, it gained another six feet and waves splashed near his shoes.

"We have to run for it." He wrapped an arm around Juliana's waist and clambering like a crab on two feet and one hand, he dragged her away from the advancing waters. The slope of the road required that he advance six feet to gain a foot in elevation.

The shaking built up again, not as violent as before, but the ground churned faster and faster. He scrambled even more quickly on a surface that did its best to thwart his efforts. A tree fell beside the road swatting them with its leaves, but its heavy branches missed them and the trunk did not block their path.

Several times the splash drenched his feet, urging him to redouble his efforts. Juliana pulled loose and scrambled ahead on her own.

Steve observed his parked pickup truck bucking and dancing around. It had slid to the side of the road and had turned 45 degrees, but its brakes held. At least it didn't roll down the road and hit us, he thought.

"Move to the side. Watch out for the pickup." Steve yelled. "I don't want to be anywhere near it right now." They scuttled on up the slope toward the top of the levee.

Weaving in and out through the trees and around the kudzu vines, Samantha and Danielle McCutchen rode their dirt bikes up the hill along the trail south of Fort Pillow. With peanut butter and jelly sandwiches in the saddlebags strapped to the back of their bikes, they headed for the picnic table near the entrance.

In prehistoric centuries, trees and vegetation covering the bluff controlled and channeled the torrential rains of the region and protected the bulk of the promontory from erosion. Only the river, chewing at the base of the ridge, had worn away the hilly remains of sand dunes left from the glaciers of 15,000 years ago. The old dunes had been transformed in time to a poorly compacted sandstone called *loess*.

During the Civil War, this vantage point high above the river became an important military position. Trenches and artillery emplacements had been put there to prevent unwanted river traffic below. The military had changed the contours and channeling of the water to fit its needs.

Different water drainage percolating through the ground developed different strains on the lay of the land.

As the girls approached the rest station at the south entrance, they felt the ground jolt, then writhe back and forth. "Ouch. My bike's falling," Samantha cried as the earth's convulsions shoved her bike to the side and threw her onto the path, scrubbing her knee in the gravel. When she tried to scramble back up, the land shook more and more, and she sprawled on her face.

Danielle stopped and stared with fright, keeping her feet spread wide on the ground, straddling the bike. Trees and vines whipped back and forth although no wind blew. "Sam, I think we must be having an earthquake."

Samantha pointed at the road. "Dan, up there. The trail is cracking apart." A growing brown fissure cut through the trail and the weeds near the road. Further along the path beneath their feet slipped down and away from the road. "The cliff is falling into the river," she screamed. "Run for it."

As they struggled up the trail towards the expanding rift, it grew larger and larger right before their eyes. The shaking worsened. By the time they could reach the rupture, the displacement had grown to seven feet, trapping them on the lower level.

The quaking grew still more violent. "The cliff, I feel it dropping," yelled Samantha as the bottom fell out from under their feet like a fast elevator. A half-mile section of the bluff, 200 feet thick and extending over 300 feet down and into the river bed, broke loose from the side of the old sand dune. Two million cubic yards of loess with trees, squirrels, bike paths, and two unwilling bike riders onboard slipped as a whole toward the river below.

Ducks burst squawking from off the water into the air, but Paula Kelly paid no attention to the birds or the growing rumble beneath her feet. She felt like crying but knew that wouldn't do any good. Somehow, she had to find a way to work through having her baby without her parents' blessings. Freddy had gone back to work and the loneliness set in. Soon she should drive back home to Kentucky.

The earth vibrated harder and growled louder, like an upset beast, like a monster about to attack. The frothing and churning waters caught her attention. Pulling herself erect with help from the park bench, she stared at the river. "What's wrong?" She turned and bolted toward her small car.

Paula covered a mere seven yards before the S-waves struck. The earth heaved back and forth, twisting her knees and forcing her to the ground. She stumbled up and staggered in the direction of her car. She must find Freddy; he would take care of her.

For over a minute the earth shook and heaved and jerked. Strange things happened to the levees and lawns around the river: they began to sink and slide into the river and the entire peninsula on which Cairo and Fort Defiance lay turned into a huge pile of liquefied mud. Some of the surface, like asphalt on the streets, held together to form a scum over the slop. Sand blows erupted at the other end of the park.

By the time Paula reached her car, the asphalt of the parking lot had broken into smaller and smaller pieces and parts of it slipped beneath the muddy water flowing up in the cracks.

Paula somehow drove the drunken car back up the road to the service station. As she stopped outside the garage, the entire building lurched to one side. The back and left walls sank into the ground and the roof fell in. She opened the car door and put one foot out to scout around the station. "Where is everyone? Where are you, Freddy? Where have you gone?" she screamed.

A sudden spasm cramped her abdomen, low and inside. Realization flooded her mind and instinct told her that the tightening of the muscles in her belly meant her body was preparing for the ageless process of birth. She screamed again as she fell from the car. She collapsed on the quivering pavement and moaned, "No, no. Not now."

"Come take a look. Don't you want to see the locks close behind the towboat? The gates are just beginning to move." Everett called to his wife, hoping she would join him. He smiled when she rose from the bench and joined him at the railing.

"The barges really fill up the lock, don't they?" she commented as she gazed into the lock basin in the Kentucky Dam holding the towboat with its nine barges.

"The engineers designed it so the barges would just fit."

"Everett, does closing the lock always make a rumbling sound like that?"

Everett too had noticed the growl. "Don't know why it should. I don't know what that is." Without warning the railing began to tremble. "Whoa, what's that? What's going on? Deb, it feels like this wall of the lock must be falling." He grabbed his wife's arm and yanked her away from the railing, running back across the sidewalk. "Get back."

The convulsions grew in intensity, and Everett could see the lock wall on the other side of the chasm next to the dam jerking even more. A truck parked on that side jumped in place then wiggled from side to side.

At that point, a full minute had passed since the earthquake began; 36 seconds after the actual fracturing of the New Madrid Fault had ended. The crack in the earth's crust pointed to the northeast, straight at the Kentucky Dam, and focused energy in that direction. The alignment of the dam paralleled that of the fault, an orientation that enhanced the resonance of the entire structure with the S-waves that rolled in from the southwest.

The earthen portion of the Kentucky Dam started to resonate with the energy collected from the seismic waves. The resonant oscillation concentrated the forces into the 600-yard long concrete structure holding the hydroelectric turbines and floodgates and shook that huge solid mass at greater and greater amplitudes relative to the land on which Everett and Deb stood.

Deb screamed. "Everett, the bridge over the lock. It's breaking into pieces." Within a few cycles of being pushed and then pulled, the bridge cracked. Then it crumbled and crashed into the walls, falling in pieces to the water below.

Everett's gaze was drawn to the huge metal gate that held back the waters of the lake as the same cyclic motion tore at it. He could see the doors twist. "It's an earthquake. The locks are twisting apart."

As the dam structure on the other side of the lock moved away to the southwest, the lock separated. Water gushed through the opening. When the dam bounced back the lock gates jammed together, wrenching the metal plates out of shape. The opening between the doors grew larger with each oscillation as the metal of the gates screamed and bent more and more.

The momentum of water bursting through the opening added to the distress of the structure, the steel of its construction already bent beyond its yield point. The barges in the lock shoved forward in the flow and pushed harder and harder against the lock doors.

Though built with ten times more strength than engineers had calculated would ever be required, the locks and their mechanisms had never been designed to withstand the combined forces that the seismic waves, moving barges, and flowing water put upon them.

As a barge hammered into the left door of the lock that had become positioned in front of the right door, the added force ripped the door's gigantic hinges from the walls of the lock, knocking a huge mass of metal and concrete into the right door. That structure, too, failed, and with a mighty roar, a 90-foot high wall of water carrying the barges and towboat blasted through the opening.

As the wall of water and its contents plunged toward the Tennessee River below, it stripped the remaining pieces of the locks from the walls of the channel.

Power from the hydroelectric generators within the concrete core had already been lost. The upper locks had closed a mere three feet, and though not damaged by the shaking, the huge motors could no longer move the locks in and out. With the entire channel open, over 45,000,000 gallons of lake water poured through the opening each second.

Everett stepped forward and held onto the railing, staring at the rushing tide. "My God, the whole lake could drain out these locks."

— 5 —

The Shaking Moves On

Buddy Joe yelled into the wheel house from where he clung to the rails above the deck of his towboat. "Jeff, full reverse props, we've got to fight this river."

"Cap'n, these waves. They're not like anything I've ever seen, even out in the Gulf in a hurricane. It's more like they're coming from underneath than from the side."

Buddy Joe put his hands onto the window frame of the wheel house. "Yeah, I feel them hitting the bottom of the boat. They must be coming from the river bottom bouncing up and down."

Buddy Joe reached inside for the radio microphone and dialed the Lady Bug frequency. "Dennis, Paul, are either of you feeling an earthquake? Are you there? *Lady Janet* or *Lady Jane*, are either of you feeling an earthquake?"

He released the microphone button and listened. A strange static came from the speakers for a moment, then a stronger signal. "This is the *Lady Janet Quayle*. What did you say, Buddy Joe? Did you say something about an earthquake?"

"Yes, Dennis, there's a big, big shaking going on, and it's tearing the hell out of everything down here at Caruthersville. The river's got a chop in it at least three feet high right now. It's doing its best to tear apart my tow, but so far everything's holding. Are you feeling anything?"

"No, it's quiet here except for the rainstorm over east. How long has this been going on?"

Buddy Joe looked at his wristwatch. "It's seconds before 9:35. I'm not sure just when it started, less than a minute ago I'd say, but it seems like it's been going on for hours. Denny, this is God-awful."

"What do you think, Buddy Joe? Is it going to shake here in Cape Girardeau? What should I do?"

"I assume it must be the New Madrid Fault that's gone off, so either you won't get hit bad up around the Cape, or it just hasn't reached you yet. Head for deeper water as quick as you can, just in case. The biggest waves are near the banks, and maybe over the sandbars."

"Okay, I'll start moving farther out. There's no traffic in the … Hey, Buddy Joe. The river's just clouded over, like there's a wind blowing. Is that what you saw?"

"Yes, it got frothy and then seconds later the waves started picking up. Sounds like the quaking has gotten to you. Hang on for dear life."

Darrell Richards, the Army Corps Traffic Officer in Memphis, spoke over the phone to John Claymore, his counterpart in the Vicksburg office, 300 river miles downstream. "John, we've got to get those towboat captains to slow down. The river is running through here at seven feet over flood stage, and we've got a ten-foot crest coming through St. Louis that'll reach Memphis by next Tuesday."

John answered, "We've got the same problem here, too. Had a report of some levee wash problems from the wake of a couple of big tows passing close by each other near Greenville. We sent a crew up that way yesterday. But it's hard to slow these barges down."

"It's happening all over; everyone's in a hurry. John, I'm especially worried about the Barfield levee up here. We keep sending out notices, but it's like the towboats don't think they make any difference."

Darrell leaned back in the old oaken swivel chair he had salvaged from his former office. His office window looked east toward the heart of Memphis, and he could see the new arena a few blocks south. It seemed strange not to face the river. He still preferred his old office where he could watch the waterway under his charge.

John asked, "Have you had any other problems? Any boils or such up your way?"

"We've added to the levee patrol and they're getting pretty good at spotting our weak spots, so I'm not worried right now. But that fast melt in Minnesota worries me. The river could crest another four feet higher if this warm weather and the rains keep up."

Darrell stretched his gimpy leg and changed the subject. "John, I don't know what you had planned there in Vicksburg, but my wife told me she had tickets for a couple of the Jazz Festival venues this afternoon here in Memphis, so I better finish up here."

Darrell heard John chuckle. "I understand, Darrell. My wife has plans to visit the new mall so I better get going, too."

Darrell sensed rather than felt a growing vibration in his chair. His voice rose. "Hey, John, something's going on." He paused and looked around. "There's a buzz in the building."

"A buzz? You been hitting the bottle this early? That ain't like you, Darrell."

"No, I'm serious. Things are starting to shake, like we're having an earthquake. I remember one we had a few years ago, but it didn't feel like this. It shook a lot …" The building jerked hard right, then left. The loose furniture slid across the tile floor. Darrell screamed, "John, it's starting to shake real hard. Everything's falling in the office and the glass panels are breaking. This must be a big earthquake. It must be really strong. Oh, God, I can't keep this chair from rolling around. John, help me."

"Darrell, what are you talking about? Are you serious?"

"John, we got a really big problem. It's a big earthquake that's hit and it's tearing up this building. God knows what it'll do to the levees. We got to …"

The filing cabinet hit Darrell from behind, knocking him into the desk and falling across his back. He screamed in agony and fell to the floor, his rib cage and backbone crushed from the blow.

Holding to the metal railing with white knuckles, Ron and Lynn watched from the second deck of the *Bella Queen* as the destruction of Caruthersville unfolded before their eyes.

"Ron, the pilings around the dock, they're pulling the casino barge in all directions. The barge is breaking apart," Lynn gasped. She could see people inside trying to pull open the doors on the casino's second deck. "Ron," she screamed. "The boat is rolling onto its side into the river. It's capsizing. Ron, those people inside can't get out."

Ron stared dumbfounded and started to speak. "Lynn, maybe they can swim downstream and …"

A huge explosion a quarter mile upstream interrupted his remark as the middle tower of the 90-foot tall concrete grain elevator erupted in flames, its sides and roof blowing apart, and chunks of concrete the size of trucks spewing into the river and back over the town. As the flames billowed out from the structure the cracked towers on either side burst open, adding more missiles up and down the riverfront.

Buddy Joe watched as the old elevators beside Caruthersville burst in a brilliant flash and black smoke poured from the pile of rubble that remained.

It took four seconds before he heard the blast from the explosion. The screaming of metal tearing against metal and the drumming of barge beating on barge filled the air with sound. He stared at the flames and smoke. It looked like the pit at the races, the car catapulting over the wall, the huge ball of exploding gasoline. Almost like in slow motion.

Reality and memory fused in his mind. "Jimmy, get down there on deck and get those guys back away from the ropes. I'll take the helm."

"Yes, sir," Jeff replied.

Buddy Joe's eyes clouded. He gripped the rudder control in a hand of steel. Why had he ever thought nothing could go wrong? Why?

Ron and Lynn worked their way around to the other side of the boat as the *Bella Queen* turned to face upriver. He placed one arm around a railing stanchion and held Lynn in the other, bracing his hip against the railing. Waves continued to race across the river surface. The boat shuddered and pitched in the water. People on the lower deck were being drenched as walls of water washed up the side of the boat.

Ron said, "Look, you can see buildings crumbling on the other side of the flood wall." Another brilliant flash lit the sky over the trees. "That must have been the power distribution station shorting out north of town. God, the place is simply being torn apart. This is worse than the bombs in Kosovo."

"Ron, look at the trees up near the grain elevators. Some of them just shook their branches lose from their trunks. And the casino dining room on the levee has crumbled." Sounds of the destruction decreased as the boat moved farther from the shore. In the distance they could hear screams and small explosions.

"Lynn, why don't we hear the earthquake?"

"Most of the sound of an earthquake comes from the destruction, not from the ground itself. If the waves and boat weren't making so much noise we could probably hear something."

The couple watched in silent horror as more and more of the town fell and disappeared behind the floodwall. In places the floodwall itself tilted to the side or slid down the slope into the river.

At last Lynn said, "Ron, I think the shaking must be stopping. The waves are getting less."

"What's that smell? It's awful." Ron wrinkled his nose as his stomach started to heave.

Lynn said, "That must be coming from all the muck stirred up from the bottom of the river. Every bit of filth and garbage that's been covered by mud on the bottom of the river in the last 100 years must have been stirred up into the water. God knows what kind of mess that is."

As the boat moved into the river current, it stayed even with town. They could see the bottom of the casino barge face up in the river with one fellow searching around on its surface, maybe looking for a way inside. Only a few people swam in the water about it.

Small streams of smoke rose from the town and turned into full black pillars with flames, but Lynn and Rob heard no sirens. "Lynn, look downstream from the casino. That must be a big break in the levee. I think water is pouring into that subdivision."

"Ron, Caruthersville has been totally destroyed. And with the flooding, those people aren't going to have any place to go." Her voice cracked. "The Captain's got to start a rescue mission for those people on shore. We've got to do something to help."

Ron choked. "You're right, but what can we do?"

Barney thumbed the handheld microphone. "Buddy Joe. I think it's quitting, at least where we are. The waves on the river are dropping, and I don't feel so much pounding on the bottom of the boat. What about you?" He waited a moment before a response came back.

"Why did I say nothing could go wrong? I told him there wasn't anything to worry about. Why?"

"Buddy Joe. Get hold of yourself. This is God's earthquake, not yours. Now, is the shaking stopping or not?"

"I'm sorry, Barney. It's just this isn't what I expected. My mind keeps getting confused." Buddy Joe's voice sounded strangled. There was a pause, and then he continued, sounding more in control. "Yes, it isn't shaking so hard up here and the river is quieting. I can see there's major destruction down your way in Caruthersville. I saw the old concrete elevators explode and it looks from here like the metal granaries up the river from town have simply split open. What damage do you see? Over."

"The casino has capsized and the flood wall is partly down. There are quite a few fires in town and any of the higher buildings still standing are missing sides or parts of their roofs and sides. All the power lines that are up look tangled, but they aren't sparking any more. People are lining up on the levee next to the river. Looks like I need to start some rescue efforts pretty quick. Over."

"Barney, Jeff pointed out a big break in the levee upstream from the *Lady Bird*. It looks like it's flooding to the west, so Hayti has a problem developing pretty quickly. The river surface is much, much calmer now. I better check damage. My barges were bouncing all over the place and some of the ties broke. Once I get things back together, I'll move downstream closer to the *Queen*. I have some flat space for survivors if you need it."

"Thanks, Buddy Joe. I think we better stick together right now. We're going to have more problems, we just don't know what."

Seismic waves had rippled through the land beneath Caruthersville and the Mississippi River for over 70 seconds. At last the quaking died out, the noise of barges clanging against each other quieted, and the waters of the river returned to a relative calm. Only then did Buddy Joe see the towboat moving upriver as it pushed in full reverse against the current of the mighty river, pulling the tow at seven knots. He cut the engine speed to match the five-knot river current.

"All hands on deck. I'm cutting the engines to half-speed reverse. We'll hold here for a few minutes then move down to just above Caruthersville." He called over the intercom, "Start checking and tightening all cables to bring the barges back together. We've got to hold this tow in one piece." He looked out as three men ran out over the barges, checking the ropes. "Where's Cecil?" he called. "Has anyone seen Cecil?"

One of the deckhands pointed downriver and yelled. Buddy Joe stepped out of the wheel house and put his hand to his ear. He heard Jeff relay the message, "Clarence said he got hit with a steel rope when the shaking first began. Took off the top of his head and threw him overboard."

"We got to go get him. Why didn't someone yell 'Man Overboard'?"

"Cap'n, he died before he hit the water. He could be two miles down the river by now. There's no way we can reach him. There ain't no hope."

Buddy Joe backed into the wheel house and sat down on the stool. His eyes streamed tears as he closed them tight. "Oh, God. Oh, God." He pounded on the front panel. Opening his eyes he looked again at the flaming grain elevators, the crumbled skyline of the town a mile downriver, the sunken levees, the twisted trees. "God, why have you done this to this land, this river, my crew? Why?"

Paula lay on the asphalt pavement of the parking lot beside the service station. "Help me. Freddy, help me," she cried, moaning from the intense labor pains and unmindful of the scrapes on her elbows and knees from the intense shaking that had stopped minutes before. The continuing small vibrations from the aftershocks went unnoticed.

"My baby's coming. Help me." Instinct told her that her baby's birth would happen now. Her uterus contracted at least once a minute, pushing its burden along. Her water had broken during the earthquake and her abdomen cramped more each time. She lay on her back with her knees in the air. Reaching between her legs, she felt the hard lump of the baby's head filling the opening of her vagina.

"Where is everyone? Freddy?" She had memories of cars careening off the bridges from Missouri and Kentucky. They had turned and rushed toward Cairo before she could reach the road and flag them down. Then parts of the bridges had fallen. After a while she saw no more cars, no people, nothing. "Help me. Please don't leave me alone."

Between contractions, Paula thought of what she could do. She recalled hearing in her Lamaze classes how the aboriginal women gave birth by squatting, letting gravity bring the newborn child out of the womb. Maybe that would help.

Reaching up, she wrapped her hand around the door handle of her car and pulled her torso erect. "Oh please, God. Help me. Help me get up." Tears streaming from her eyes, she felt as if her entire abdomen would tear apart. She struggled to gain the leverage she needed.

She pulled a knee under her thigh then lifted to get her foot into position. She could not feel the pain of the gravel on the pavement cutting into her leg, it did not compare to the pain of labor.

Taking a deep breath Paula heaved her body upward and brought her other leg around. At last, she squatted, hanging onto the door handle of the car. "Thank you, Lord. Thank you." She dropped her head on her chest and pushed to balance the pressure in her lower body.

Reaching down under her dress, she tore her panties to the side, ripping them aside to leave room for the infant emerging from her womb. She shifted her feet farther apart, ready for what must happen. Now she could have her baby.

Holding her hand under the head emerging from her vagina, she held her breath and heaved again and again. A great flood of relief enveloped her body as the small newborn slid down the birth canal and dropped to the pavement below. Paula leaned against the side of the car and reached down with both hands to bring her infant up in front of her face. The umbilical cord trailed down its belly, still connected to the placenta left in her uterus.

"My baby. My little darling." Paula grabbed the hem of her dress and pulled the cloth up to wipe the mucus clear from the face of the infant, cleaning its eyes and nose and mouth. She wiped its stomach then turned it face down and cleaned its back. Patting it on the back, she rubbed its ribs, trying to stimulate it to breathe, to take a breath. She waited, and began to shake in fear.

The baby convulsed and expanded its chest, sucking in a mighty gasp of air. With a prodigious wail, it announced its arrival to the world. Again and again the newborn cried from the sudden realization that it no longer rested secure and in the warmth of its mother's womb. As the baby cried, the earth again shook in chorus.

Paula clutched her baby to her breast and cried with it, sobbing with relief and joy, bonding with her child like mothers of aboriginal times. She held the baby out in front of her and looked. "You're a girl. I have a baby daughter." She cradled the wailing infant in her arm, still holding onto the car handle.

For several minutes Paula squatted by the car. Her abdomen again cramped. It felt like she must give birth a second time. Then she felt the reward of her pushing as the placenta dropped to the pavement.

Paula kicked against the remains of the birth, trying to find some way to disengage the umbilical cord that bound her baby to the bloody mass at her feet. "Oh my daughter, you're still tied to that thing. I've got to get you loose."

Finding nothing to help, she pulled the cord to her teeth and gnawed it through. Struggling, she stood, leaning against the car. The baby cried and she held it to her body, one bloody hand wrapped around the end of the cord in an urgent attempt to stop the flow of fluids, unaware that evolution long ago had solved the problem of the loss of blood from a newborn.

— 6 —
A New Reality

At the south end of the Three Rivers Wildlife Management Area in Louisiana, water flows from the Mississippi River through the Army Corps of Engineer's Old River Control Structure. There it joins the Red River and forms the Achafalaya River, a "distributary" of a Federally mandated 30-percent share of the Mississippi River's water flow.

Boy Scout Leader Guy Long pointed to the massive concrete dam alongside the river and explained its workings to his troop of eight middle-school-age boys. They were on a spring camping trip from Natchez, Mississippi, 45 to the miles north by way of the river road.

"Here you see where the Army Corps of Engineers have built a great concrete dam reaching over 100 feet down into the mud to hold the Mississippi in place. Only part of the big river's water is allowed to flow into the Achafalaya River basin to keep the swamps flooded and support the wildlife. If the big river had its way, all of its water would flow into the old channel and Morgan City would be the major seaport on the Gulf Coast instead of New Orleans."

His smallest charge, a thin youth with thick glasses, asked, "Mister Long, what holds the dam in place?"

Guy smiled. Henry always asked the probing questions. "I read that dams mostly depend on their weight to hold them down and keep them from moving. They remain where they are built because of the friction between their base and the solid rock underneath. But I also read that the Army Corps of Engineers drove steel pilings into the river bottom, just in case, to hold this dam where it is."

In the Mississippi River basin there is no native bedrock, only semi-firm old river mud and sand saturated with ground water. Guy Long did not mention that in 1973 a massive flood had almost undermined the Old River Control Structure, nor was he aware that today the river ran high, just a foot below flood stage, and once again eroded the river bottom at the foot of the structure.

Two minutes and 40 seconds after the 70-mile fracture of the New Madrid fault had halted, a burst of seismic energy swept through the swampy flood basin. Guy identified the funny feeling first and called out. "Your attention, boys. Feel that movement? Somewhere, but not very close, there has been an earthquake, and we are feeling it shake back and forth right here. It isn't very often you get to feel an earthquake, is it?"

The rocking motion grew to a steady swell and continued. He frowned. "It sure seems to be shaking for a long time." For 75 seconds the region lurched back and forth once a second with forces between one-tenth and two-tenths the acceleration of gravity.

His troop laughed and Guy worried as the group ran, stumbling across the dusty gravel of the parking lot back to his purple van. The intensity decreased, but a subtle sloshing motion continued for another minute, enough to keep the sand and mud in suspension in the liquefied soup that had been created beneath the old river channel.

Henry, the boy with the glasses stopped, and stared. "Look, our bridge to the other side just broke." One by one the others came to a halt. The scouts stood in the parking lot with their leader and watched the slow-motion migration of the Old River Control Structure, like a turtle extending its head out of its shell. With no friction left to hold the concrete dam in place, the hydraulic forces of the big river simply pushed the dam aside as gravity buried it in the slush, pulling the steel pilings out of the liquid bottom. Now the big river flowed the way it pleased.

The victory of the Mississippi River that many called inevitable had happened, not the result of a flood but of an earthquake. Baton Rouge and New Orleans would soon sit beside a stagnant slough, cut off from the main river channel and the life-giving flow of water needed to fill their shipping channels and support their factories and refineries.

On the other hand, Morgan City did not have a waterway large enough to handle the huge volume of water headed its way. In time, the flood would gouge a new course for the river and begin a new delta into the Gulf of Mexico. Along the way it would destroy all the roads, bridges, railways, canals, power lines, and pipelines connecting Texas and west Louisiana to the Gulf Coast states in the east.

Guy Long sat in the driver's seat of the van. As he watched the accelerating flow of muddy water into the outflow channel, it felt as if his world had come to an end, just like the road back to Natchez. He exclaimed to the troop. "Kids, I'm not sure how or when we're going to get home."

Most of the stunned passengers of the *Bella Queen* stood next to the lifeboat stations on the south side of the boat, peering across the water at the remains of Caruthersville just 300 yards away. A few gathered on the other side, not wanting to view the catastrophe, preferring instead to gaze upon the trees lining the opposite bank. The waters had quieted. Everyone on board seemed safe as the boat held steady near the middle of the river.

Many had watched the grain elevators and water towers at Caruthersville do their crazy dance, then explode or crumble to the ground out of sight behind the trees or the levee. They had seen the casino boat, which had been tied next to where they docked last night, jerk about and tear apart its ropes to the bank even as it broke in two and capsized. Pillars of black smoke rose throughout the town behind the floodwall. A growing horde of people crowded onto the fallen remains of the floodwall as river water lapped at their feet. They had seen the town on shore die.

"Buddy Joe, this was the Big One." The Captain of the *Bella Queen* spoke over the ship-to-ship radio. "I never thought I'd see the day, but this looks like the end of the world."

"Barney, I know what you mean. I lost a man who was hit with a cable, but I think the people on shore must be far worse off. What about you?"

"So far as I know we didn't lose anybody, but you're right. I can see people gathering on the levee. As soon as it's clear I'll organize rescue teams to go in and pick up survivors."

Buddy Joe said, "I can now see at least a quarter mile of levee up at Gayoso Bend has settled, and there's a big crevasse there. I just saw a tree fall over and flow out the channel. There's a flood starting west of here that will take out Hayti for sure." He kept his towboat steady, holding even with the grain elevators above the town.

"Did you tell anyone?"

"I called the emergency channel and reported the levee break, but they said they were busy with other reports. At least they know."

Buddy Joe continued, "Barney, I scanned the channels for radio traffic upriver and told a few of them a quake was happening. Everyone but Denny Bugler laughed at me, but I warned them. I only got a few reports from downriver, and those seemed garbled, like everyone was too busy to return. The shaking seems to have moved from down south to the north. That agrees with what I see of the maps of the New Madrid Fault."

Barney thumbed the mike to respond. "Yeah, I listened to some radio traffic but couldn't make sense of what was going on. What do you think? Do we just stay here?"

"Yeah, let's hold here for the time being. The chop seems way down, so maybe the shaking's stopped. We need to talk to people on the radio to find out about the situation in other places. We aren't even sure where the earthquake happened." Buddy Joe looked at his watch. It read 9:39, only five minutes had passed since the river went crazy.

"Oh my aching back." From time to time Virgil leaned back from his task of bailing water from his boat with his old bait bucket to rest his bones and look up. I can't believe that whole bridge fell down, he thought. In place of the southern half of the highest part of the bridge he saw an expansive view of the sky.

Virgil remembered hanging onto the rebar as the world around him went to hell, seeing the bridge break up and the long piece of concrete fall, creating a huge splash as it disappeared beneath the water. The feel of the water surge from the falling slab that washed up and over his concrete support flashed across his mind, and he thought, I sure had to hang on tight or that flood would have washed me away. He spoke to the heavens. "Thank you, Lord, for leaving my boat tied to the column and not sunk." He had found it half full of water, tugging on its rope as it drifted back and forth in the swirling current.

As Virgil dumped another bucket of water out of the boat he felt tugging under his foot. Looking down, he saw that his boot rested on the handle of his fishing rod. The tight line running into the water whipped the tip back and forth. He reached down and took his fishing gear in hand. As he raised the rod, he felt a tremendous pull on the line and the brake slipped to allow more line to pay out from the reel.

"Whoa, now that feel's like a big'un." Virgil tightened down on the brake and started to reel the line back in. He could feel a big fish jerking at the line, trying to swim back toward the bottom. Adrenalin shot into his aching body, erasing the pains. "Come on, Grandpa, I gotcha' this time."

Virgil fought the fish up, then down, then up and down again for several minutes. Both he and the fish tired. Then another aftershock pounded the river and the mud around the column where he had tied his boat. The line went slack. "Oh no, I done lost my fish."

When the shaking stopped Virgil reeled the loose line back onto the spool. "Old Grandpa fish, I shore would've liked to at least seen you." He had wound most of the line back onto the reel when he noticed the remaining line heading upriver and coming near the surface.

Four six-inch whiskers preceded the head of the big fish as it swam to the top of the water near the base of the column, gasping for air. Virgil crouched in the rocking boat for a better look. "Grandpa, you're at least six feet long and you must weigh over a hunnerd pounds. You're the biggest catfish I ever seen, and it seems to me you're just eager to leave the river."

Virgil smiled, a fresh supply of adrenaline coursing through his body. "Now, just you wait right there, Mr. Catfish. I know you're afraid of this shaking water about as much as I am, so I'm just going to get my boat around here so's I can land you, and then we'll finish this fishing business." He held the rod in his left hand while he tugged on the rope to bring the boat nearer the column.

The bow of the boat scraped along the concrete, making a bumping noise, and the startled catfish turned and dove for the bottom. Virgil kept his left thumb on the reel as the catfish stripped out the line. The moving force almost jerked him out of the boat, but he plopped down next to the cross-board into the water still puddled in the boat bottom. He reached with his right hand to take the handle of the rod and reel.

The fish continued to pull line from the reel. Smoke poured from under Virgil's thumb, and as he changed hands, he looked at his appendage in awe. The pad of his left thumb had been rubbed raw. It turned deep red and hurt like hell from the heat. He reached over the side and swished his hand in the water.

Virgil tightened the brake and slowed the fish and then started reeling him back in. With patience and care he fought the huge fish back to the surface. Then, taking the gaff from the bottom of the boat, he raised to his knees and reached over the side to hook it under the fish's chin. Dropping the fishing rod into the bottom of the boat, he took the gaff handle in both hands and leaned back, struggling to slide the flopping weight into the boat.

The fish threw its head from side to side, almost jumping into the boat. Falling on the other side of the cross-board from Virgil, the fish flopped from side to side in the bow of the boat as Virgil sat back down in the water-filled bottom and looked at the biggest fish he had ever caught.

"Boy, will I ever have something to talk about at supper tonight."

When the convulsions from the earthquake ended in Caruthersville, Loretta stopped running from the river. She had dodged sand blows and seven blocks of broken and falling buildings, as she trotted barefoot in the middle of the water-filled street past the center of town.

"Help me, please. I can't stand." An old woman crawled out to the sidewalk in front of her broken house, dragging her limp leg along behind her.

Loretta stepped out of the street and knelt beside the old woman. "What happened?"

"I ran out of my house when the shaking started. Then the porch post pushed me sideways as I stepped off and I twisted my leg and thigh. I need someone to call 911 for me so I can get to the hospital."

"I think you better stay right where you are." Loretta pulled the woman to the side. "Here, lean up against your fence and wait. All the 911 people are busy right now, but they'll come by as soon as they can. Okay?"

"Thank you, dearie. I'll wait if you say so, but if you see someone that looks like a doctor or nurse, please send them down this way." Her wrinkled face broke into a pale sweat.

Loretta nodded her assent and rose to continue her walk away from the river. She came to another house whose roof had fallen. She joined others who were digging a screaming couple out from under the broken timbers.

Small aftershocks continued to pound the area. Some were strong enough to shift the rubble, but most simply added to the feeling of dizziness that everyone felt from the unstable ground.

The first significant aftershock, a 5.7 magnitude event, struck 20 minutes after the big event with an epicenter 8.5 kilometers beneath Caruthersville. For four seconds the town shook from seismic waves at a quarter the force of gravity. After 30 seconds, gentler reflected waves shook the town apart once again.

Loretta panicked when the shaking began but became calm when it ended. Several buildings down the street completed their fall, but many unstable walls still stood.

Other shocks coursed through the town, but some people ignored them and focused on rescuing those who were still alive. Loretta joined in trying to find the injured and remove them from the ruins. The dead they left where they lay.

"Come on, kids," the young father said to his dirty seven and nine year-olds. "Let's go to the levee out by the river with the other people. It's the highest place around, and I don't want the river to catch us down here in town." Turning to Loretta he asked, "Don't you want to go with us, ma'am?"

"No, I've got to help get these people free from the ruins. They won't have a chance if we don't pull them free before the water gets here." She returned to her task, prying with a broken fence post to lift the fallen roof of a house so her helper could look inside. Her yellow dress had been torn and smeared with mud and her black hair hung in a tangle to the side of her head.

More and more people in Caruthersville who were able hurried north and climbed onto the levee next to the metal granary a quarter mile above town. That levee had not slumped like the one protecting the town, and soon more than 400 souls clustered together on its wide crest. Water from breaks upriver started flowing across the roads in front of the levee. This created an island of higher land on which the survivors could seek refuge. Now, even more people waded across the growing lake to reach its safety.

The floodwall next to the town held back the river, but water now splashed a foot up the wall. The flow through the opening at Front Street had been quenched with sandbags. People began putting barriers in the streets to the south and east, trying to stop the advancing floodwaters from crevasses downriver. In time, hand-built dikes and other barriers prevented the water from covering the town, creating the edges of another island with its high point at the center of town around the courthouse.

"Help me," Loretta called to people hurrying by. "We've got to get the injured people over to the courthouse." She supported the old woman hopping on one foot along the street.

"The courthouse is not good enough. We're going to the levee. Our neighbor said it's 10 feet higher than the town." They rushed on.

"Paula, Paula! Are you okay?" Freddy ran from the direction of Cairo. "I followed the cars heading toward town, then I remembered you hadn't left the park." His breaths came in gasps and he scrunched over from an ache in his side.

Paula leaned against the car and smiled, tears streaming down her face. "Freddy, see our baby. I had our baby. All by myself." She held the tiny girl out with pride for Freddy to see.

He stopped and stared. "Is that our baby? It looks like a doll. Does that mean I'm a father?"

Another aftershock rocked the ground. More trees fell. The land quivered like gelatin and seemed ready to flow.

Freddy searched around. "Paula, we've got do something, get to a barge. The land is sinking out from under us." He grabbed the suitcase and blankets from the car and helped Paula. With the baby in her arms, she and Freddy ran for the barges tied on the side of the Ohio River. After they had climbed onto the first one they reached, the intensity of the quaking from the aftershocks became stronger and the land they had left dropped lower.

Ron and Lynn worked in the lifeboats at Caruthersville with other volunteers and the crew to bring 412 men, women, and children and three small dogs on board the *Bella Queen*.

Finally, no more people climbed onto the levee. Others who could climbed to the tops of the trees or the broken buildings, or they fled inland or to the levee. The Captain's concern about the increased water flow through the crevasses convinced him that it made good sense to move back to the middle of the river.

Ron rinsed his hands in the bucket of river water sitting on the deck. "You know, this is as bad as some of the things I saw in the Balkans. But at the same time, seeing the suffering there makes it easier to handle this tragedy."

Lynn looked at him. She had a look of reproach in her eye. "I suppose, but I don't believe I can ever wipe the memory of those people crying for help on the levee."

Ron and Lynn walked back to their cabin. Opening the door they found the small room occupied by an older couple in their sixties and two small children, both under four. All were still muddy from their climb over the levee wall and into the lifeboat. The children sat on the upper bunk dangling their muddy shoes in front of the adults on the bottom.

The woman started to rise from the bunk. "The steward let us in this cabin so we could rest. We didn't touch anything. I'm sorry, we got the bed a little muddy."

Ron remembered the family from one of the trips between the levee and the riverboat. Lynn reached out to touch the woman's shoulder. "That's okay, you don't have to get up. Stay where you are."

The old man held out his hand to Ron. "We're Tim and Alta Warren. These are our grandchildren, Timothy and Ruth." The children smiled and looked down.

Tim leaned forward and looked up at Ron, staring through bloodshot eyes. "We were in the buffet room onshore at the casino when the earthquake hit. We got out just in time. The whole building fell into the river. It was awful."

He looked down at his dirty hands and shuddered. "Our son and daughter-in-law, they were in the casino, on the boat. We watched it break apart. It rolled over into the river, and then it ... only a few people walking around the outer deck made it off the boat into the … it looked like the casino doors jammed. Everyone inside when the boat capsized was trapped and … we watched for a long time but nobody ever came up from … all just stayed down there … in the boat." His head bowed lower and lower as he spoke. His shoulders shook from sobs deep within his chest.

"Grandma, where's momma and daddy?" Alta rose and hugged her grandchildren to her breast and buried her face in their tangled hair.

"I don't know, Timothy, but I think they're already in heaven."

Ron stood ill at ease at the door of the cabin. Lynn cried with the couple, then stepped over and put her arms around Alta. "Oh, that's so terrible. We're so sad for you. But at least you're okay, and the kids are okay."

"I know. I'm thankful I have the babies, but I've lost my son and daughter-in-law."

Ron rubbed the older man on the shoulder. "We're Ron Cannon and Lynn Browne. You stay here for as long as you like. We'll figure out how to double up until the Captain can find a safe place for us to dock."

Tim shook his hand. "Thank you, and I hope your loved ones are okay."

Ron abruptly remembered his own family. "Oh, God. My parents had reservations at the Peabody Hotel in Memphis for the Jazz Festival. Lynn, you and I invited them out to supper this evening, remember. If it shook this bad here, could it be as bad in Memphis? Where were they?" His eyes began to cloud. "And my girls in Nashville with my ex, have they been hurt? And Lynn, your daughter staying up near St. Louis. Do you think she's okay?"

The shock of what had happened around him exploded in Ron's mind. For the first time he realized how much of his own world had been turned upside down, how the Balkans had been so different.

Steve and Juliana's frantic contest to stay atop the levee lasted 72 seconds before the shaking eased. The waters of the river still tossed and churned, but the river ended its advance. The front wheels of the truck sat in a foot and a half of water as the jolting died down and then stopped.

Steve's entire body shook from fear. He wondered what would happen next. Standing, he drew his daughter next to him, pulled her shoulders tight into his chest, and gasped, hoping to slow the pace of the adrenaline coursing through his body. Juliana sobbed into his dirty shirt.

"Honey, I think it's stopped, but I have to save the truck. Stay here while I back it up to the top of this levee." He released Juliana and ran down the slope to his truck.

Wading into the water, Steve opened the door and stepped in. He turned on the ignition and cried with joy when he heard the engine turn over and fire. Putting it into reverse, he backed the pickup with care to the level road atop the levee.

Juliana climbed into the truck. Steve wrapped her in his arms and they stared out the window in wonder. They could see river water within four feet of their road. A few minutes before they had been unable to see the water through the trees.

"Daddy, the river can't rise that much that fast, can it? It just doesn't happen."

The awful truth dawned on Steve. "No, the river hasn't changed, it's the land. The levee's sunk, it's sagged. Thank God it didn't drop any farther. If it had cut through here it would make a huge rapids and waterfall. It would have been catastrophic."

— 7 —
A Pause

Charlie scrutinized the river from the helm of the *Amanda Blair*, searching for anything that could tell him if the crisis had ended.

He remembered hearing a low growl, like a freight train crossing a trestle. The sound grew louder and came from every direction. Then he felt his fishing line vibrating, more than just singing in the current, and he saw the brown water mask over like the surface had been frostbitten.

The first real indication that something was terribly wrong came when his taut fishing line went slack and then jerked back, bending the tip of his fishing rod over two feet before the line went slack again. Something big moved his weight on the river bottom back and forth, faster and faster.

He felt the boat jerk as the anchor rope mimicked his fishing line, tightening and slacking. On the quiet of the river, the growl grew to a roar and the water began to boil and froth. He stumbled as he stood up from his deck chair on the fantail deck and started reeling his line in as fast as he could.

He remembered calling to Sylvie. "Get up here. Something's wrong, Sylvie." As she stepped out of the stateroom, he could see her spread her feet like a good sailor and clutch at the door jam to steady herself against the growing pitch and yaw of the boat.

"It's the big earthquake they've been warning us about," he yelled. "The New Madrid must have broken loose. Hold on and stay in the center of the boat." As the sinker reached the tip of his rod he dropped his fishing gear to the deck and staggered up to the pilothouse to flip the anchor switch to RAISE. He did not want to have any rope or chain tying his boat to the earth.

For over a minute the river remained in total control, tossing the boat about in the waves. Although the boat floated freely in the eddy below Chickasaw Bluffs Number One, but went nowhere. Charlie started the engine but left the props in idle, unwilling to attempt any evasive action. He could think of no place to go.

At last he could breathe a little easier. The water no longer frothed and the chop no longer pounded the boat from all sides.

Sylvie walked to the deck chair in the middle of the fantail. She screamed. "Charlie, up there in the trees. The bluff started to slide but only came part way down. Those are the girls we saw. They're trapped."

Charlie put his head out of the pilothouse and looked up. "Sylvie, we just had a monster earthquake. This whole land has been shaken to its core. We're just lucky that bluff didn't come right down on top of us."

"But those girls, what can we do?"

"We can't do anything. They're on the side of a bluff 150 feet above us. I agree they're trapped, but someone will come along and pull them up to the top. We can't do anything from down here on the river." He shifted the props into gear. "We need to find someplace where people know what's going on. I think we better head for Osceola."

"We can't leave those girls there all alone on the side of that hill."

"Sylvie, we don't have any choice."

She dropped to the deck chair and clutched at her chest. She sobbed into the binoculars. Charlie's eyes stung with tears as he gunned the engines and swung the boat around to the north, away from the bluffs. He could think of nothing to save those poor girls; they would have to find safety for themselves. He knew that Sylvie must feel that she was again losing her babies. Once again he had failed her.

"What are we going to do now, Everett?" Deb leaned against her husband's arm, staring at the remains of the I-24 Bridge across the Tennessee River a mile below the Kentucky Dam. "Isn't there some other way to get back to our trailer?"

The roar of the water rushing through the broken locks a mile away almost obscured his answer. "… no way to the other side. The road across the dam is gone and I-24 is gone. Besides, I don't think I would want to be on that side of the river. It looks from here like our campground is okay, but everything downriver, and probably most of Paducah, must be flooding by now."

"But Ruggles is alone in the trailer. He'll starve."

"Deb, I'm afraid we've lost our trailer and everything in it, including your cat."

"Can we call someone?"

"They've got bigger things to worry about, and besides, the cell phone doesn't work. I tried that a while ago." He put his arm around her shoulder. "Deb, we are among the lucky ones. We're still alive and not injured. There must be thousands hurt. We need to get away from here as soon as we can. We have the truck and I filled it with fuel last night, so we can go somewhere that's safe, but not back across the river."

"But where is that? The radio isn't working. We know there was an earthquake, but we don't know where."

He thought a moment. "Let's head east on I-24. Maybe we can find someone who has some news. Maybe they will know something in Nashville."

Paula and Freddy walked along the side decks of the empty barges tied beside the sunken remains of Fort Defiance, searching for one with some kind of shelter. "How about this one? It has this little shed where they stored the tools, and it's pretty clean." Freddy held the door open to the storage closet on the topside of the barge.

Paula lifted her smudged skirt and stepped into the tiny room, only two by three feet in size and four feet high. A space in the corner allowed her to sit down on the steel plate and lean against the warm sheet metal siding.

"This will do just fine, Freddy. The baby and I can fit in here just great, but there may not be much room for you if it really starts to rain. Is that okay?"

Freddy smiled, and bowed from the waist. "My lady, my only concern is a place for you. I can take any of the storms that the elements may inflict upon me. Fear not. I am devoting myself to the safety of you and our daughter." With a big smile he removed his baseball cap and swept his arm down and out.

Paula bowed her head and snuggled deeper into the small room. "Oh, Freddy, you are so gracious. I don't know how I will ever make it up to you, finding this wonderful place for us to stay."

Freddy squatted on the barge deck just outside the tool shed and looked in. "Do you really think this will be okay? It's awfully small."

Paula beamed, happy with her family and the attention. "With you around, Freddy, anything is okay. At least here we're safe from the river waters."

The historical figures in the murals on the floodwall protecting Old Town Paducah looked out across the chaos of the crafts bazaar filling the parking lot. The lot had become a jumble of merchandise, cloth, and tent poles filled with the bazaar's dazed patrons. The renovated buildings and boutique shops surrounding the open-air market had been reduced to piles of brick, remnants of walls, broken timbers, and screaming victims.

Shelba poked at the bits of crafts supplies, which lay strewn on the asphalt, unsure if they belonged to her or to the woman across the aisle. "George, I'll clean up here around our booth. You go help search in the buildings."

George knew better. "I'm not going near those buildings. Some pieces still standing are teetering. If it shakes hard again they'll fall for sure." He knelt beside his wife to help. "Besides, there are enough people working over there. I'd only get in the way."

Along with several others, the couple packed merchandise into boxes, placed poles that had supported the shade cloths back in an upright position, and tried to make order out of what had been a happy and bustling fair only 20 minutes ago.

The siren of a parked police car a block away split the air. A policeman climbed into the back of a nearby pickup and shouted through a bullhorn, "Attention. Attention, everyone. Listen up. I just heard by radio that the Kentucky Dam's leaking. There's a wave of water coming down the Tennessee River and it's flooding the Ohio. We need a crew to block the gates in the floodwall with sandbags or whatever we can find, else all of Old Town's going to be flooded. Do you hear me? The dam's broke and there's a flood coming. We've got to block the openings in the flood wall."

George rose from the ground. "That's something I can do. I've worked levee patrol before. But, Shelba, if what he says is true it may not be enough. You pick up whatever you can real quick and get it to the van. I'm not sure the floodwalls are high enough to protect us, and we may need to get out of here as soon as we can. So hurry. I'll meet you at the van if we need to leave."

Shelba looked up, the strain showing on her face. George heard her admonition, "You be careful, George. I love you."

Jud sat in his johnboat with his sister. "Here, have another beer. The trees have quit shaking and the water has stopped sloshing, but I think we're better off on the water than on land for a while. At least 'til we're sure things have settled down." He popped the top off the longneck bottle.

Sally Mae took the proffered bottle. "Was that really an earthquake, Jud?"

"That's all I kin think it'd be. It tore hell out of things. Good thing you was at the dock 'fore it hit. Else you'd have been caught in the house when it fell in. I s'pose we kin get our stuff out of that pile of kindling any time we want."

Jud's johnboat was tied to the remains of the small dock on Mud Lake, the ox-bow lake that had been created when the Army Corps of Engineers had dredged the Driver Cutoff channel below the Chickasaw Bluffs Number 1, trying to improve navigation through the Plum Point Reach.

"What are we going to do now? Since ma died this place on Mud Lake has been all we had. Now we ain't got nothing. And you won't work. You just do your meth and steal things for Jake. Seems like we've hit the bottom, what with this earthquake and all." She took a swig of the beer and pushed a wisp of stringy blond hair away from her face with the back of the hand holding the bottle.

Jud scratched his itchy knee and rubbed his pimply chin. "Seems to me this might be the time to get some really good stuff. I bet ever' body is all shook up and confused. We kin just go in and collect whatever we need while they're not looking. They'll all be so busy trying to fix things up they won't pay much mind." He took another pull on his bottle. "What do you think? Want to help me?"

"Well, we got to do something. I s'pose so. Where do you think we should go, somewhere around Osceola?"

Jud grinned at his sister. "No, I think we got a great chance closer than that. We kin go after that big cabin cruiser that's down there off Fulton. They were anchored just outside where our lake connects to the river, so it's only a couple of miles. I bet they're still hanging around, and Jake said he could find a place to sell that big boat if I got it. And we could live in it while we're waiting for him to find a buyer. What do you think about that?"

Sally Mae gazed into the distance as if lost in thought, then nodded. "Okay."

Jud shifted his body on the seat board. "There'll be one problem. We'll have to dump the old man and his wife into the river. That work with you?"

Sally Mae wrinkled her brow. "Won't people suspect something?"

"I figure there'll be bodies floating down this river for days from the earthquake. Nobody'll know where they came from, or why."

"Yeah. In that case I s'pose so." A half smile flowed across her face. "It'd sure feel good to sleep on a nice big boat. And maybe the old woman will have some nice clothes."

The *Lady Bird Jamison* and its barges continued to hold stationary in the river downstream from Caruthersville. The *Bella Queen* remained a quarter mile downriver from the towboat.

Buddy Joe thumbed the mike and answered back to the call from his sister towboat, the *Lady Janet Quayle*, at Cape Girardeau. "Thanks for the update, Denny. I'll pass the word to the folks downriver."

Buddy Joe adjusted the frequency on his ship-to-ship radio and called the *Lady Jane Wilson*, his sister ship downriver near 0sceola. "This is Captain Buddy Joe Simpson on the *Lady Bird Jamison* calling Captain Paul Taylor on the *Lady Jane Wilson*. Paul, are you hearing me?" Buddy Joe waited for several seconds before he got a response.

"Hey there, Buddy Joe. I'm just holding steady down here in the middle of the shipping channel like you suggested. I've pulled about 15 refugees out of the water. I can see Osceola over the trees, and they must be having some real problems there because there's smoke rising all over the place. It looks like all the grain elevators either exploded or collapsed. I haven't ventured near shore yet because there is still some fire on the water. Over."

"Yeah. Sounds like what I'm seeing up here in Caruthersville. Paul, what I called about is a report I just got from upriver. The locks at the Kentucky Dam broke in the earthquake and the water's flowing full throttle through the lock channel. They told me it's flowing 40 or 50 million gallons a second. That's three times what the Mississippi normally carries when it's not flooding.

"The Ohio River is rising faster than they can handle according to the radio report. It's expected to hit 15 feet or more at Paducah before noon. The surge will cover Cairo by three o'clock this afternoon, and they expect a minimum of a 12-foot rise on the Mississippi River on top of the current flood levels at the junction of the Ohio and Mississippi. You copy all that so far?"

Buddy Joe listened to Paul's initial reaction. "God, that could take out all the upper delta. Now's their chance to use the Birds Point-New Madrid Floodway."

The Floodway, begun after the devastating 1927 floods, is 200,000 acres of farmland that can become a flood channel running on the west side of the Mississippi from just below Cairo to just above New Madrid. To relieve the pressure of a flood crest on the upper delta the Army Corps must use explosives tó blow the levees at designated "fuseplugs," creating crevasses through which the river water can flow.

Buddy Joe laughed wryly. "There was some radio chatter on that. It seems all the bridges are gone into the area so there's no way for them to get out there and pump the binary explosives into the pipes to blow a crevasse. But nature has taken care of that. There are crevasses all over the place, just not where they planned. But there's still too much water for the flooded lands to handle."

He continued, "The Army Corps has concluded that there'll be a big push of water to the west into the Missouri flatlands all the way over to Crowley's Ridge, but there's still a lot of water to come down the Mississippi. We're going to have added flooding all the way to New Orleans. Every levee downriver is expected to fail, if it isn't already down. Over."

Paul took a moment before he answered. "Oh Lord. Any extra foot is going to make a real mess down here. I already see a couple of places where it looks like levees alongside Osceola must have breached into Arkansas. When do we expect the surge to hit here?"

"Paul, the water rise is expected to begin in Caruthersville after midnight and hit Osceola by nine tomorrow morning. Watch out that you're not too close to a levee when the surge comes through, else you could get sucked into the break."

"Okay, I read you, Buddy Joe. We'll keep our eyes open and please give us warning if you hear anything else. Call when you see the surge come through your area."

"I sure will. And Paul, would you call around Osceola and farther south to let Memphis know that they should expect the surge in their port starting about noon. Now I better go tell the other folks in this area about what's coming their way."

Buddy Joe again adjusted the frequency of his radio and prepared to call Barney on the *Bella Queen*.

By the time the shaking had ended, the river washed over portions of State Route 22 south of Bessie's Neck. Steve Pauli walked along the road taking photos of the pavement and nearby sand blows. What he saw made him less and less confident about driving through the water to escape southward to Tiptonville. The continuing small-to-moderate aftershocks seemed to be causing further settling of the levee on which his pickup rested.

He listened to the small hand-held marine radio he carried in his truck. Amidst the static he heard reports of several levee breaks along the river. The river radio traffic seemed confused and uncertain about just what had happened.

"I think we better move," he said to Juliana.

"Daddy, why can't we just stay here? We're higher than the river."

"I don't think we're in as safe a place as we should be right now. We need to get away from where the river runs so close to itself, but the road south doesn't look safe. There's water coming from somewhere and running over the road."

"Can we turn around and go back to the highway? I saw on the map it goes north."

Steve shook his head. "The highway heads west for half a mile and drops below the level of the river before we could turn north," he told

Juliana. "With the water so eager to come over the top, we'd be tempting fate too much."

"Where else is there to go?"

Steve talked his decision through and made up his mind. "Honey, I believe our best choice is to stay on this levee road and head toward Bessie. It's a mile directly north on this dirt road. There's not much there anymore, but it should be a little higher than where we are. That's where we'll go." He started the engine of the old pickup and moved along the levee road in low gear.

"Daddy, the water's so close to the top of the levee. Don't you think driving on the dirt road might cause it to break up?"

"I'm going really slow to keep that from happening," he said and he watched the water to his right with suspicion. "I think we're okay for now." He prayed a silent prayer he could make it as far as Bessie.

Virgil pulled his boat around to the other side of the bridge pier to get into the shade. The huge catfish in the bottom of his boat still twitched from time to time, even after Virgil had driven his bait knife into the middle of the 11-inch wide head to pierce its small brain.

"Now, just how am I going to get this fish home?" He could see the rear bumper of his old pickup sticking out of the dirt from under the edge of the bridge section that had fallen on it. "Ain't going to drive that truck no more. I don't even know if the road still goes back to the highway. It's almost like the Lord don't want me to leave."

He scanned the bank where he had launched his boat. "I'd swear the river's dropped a foot since this morning. But that don't make no sense, cause the pier's a foot under water now. It had two feet of freeboard when I got here. I figure the bridge must have sunk in the earthquake." Indeed, the huge bridge columns had sunk another three feet into the Mississippi bottom in which they were buried.

Virgil heard a rumbling noise from the remains of the bridge above him. It had been quiet for the past hour. "They must've reopened the northbound lanes, 'cause that shore sounds like a small truck. I kin see the southbound lanes are all down."

Virgil used an oar to push against the top of the pier, trying to move his boat closer to the upstream column where there was more shade. As he leaned back and looked through the gaping hole left by the fallen section of roadway, he saw puffy white clouds begin to fill the blue sky. "I expect those little clouds will grow into thunderheads this afternoon. Seems like that always happens this time of year."

High above at the center railing of the northbound lane a head appeared and a person leaned out and over to look down.

"Now who's that? Must be someone up there checking the bridge." Virgil waved and the man waved back. "Maybe he'll send someone down here for me, but it don't make no sense that he'd do that. Lord, you still haven't told me why I'm here."

First George helped fill the sandbags and then he worked at the other end of the line, dropping the sandbags into the opening in the floodwall. The brightly colored, happy murals that depicted the history of Paducah covered the eight-foot concrete wall on either side of the opening. The paintings provided sharp contrast to the tense, hurried pace of the men and women building their defenses against the river.

George placed another bag on top of the growing pile. He looked at the water on the bank just outside the barrier. "Hey, that water's coming up awfully fast." He stood and stared. Even as he watched he could see the level rising faster and faster.

"The river's coming up too fast for me. I'm out of here." He turned and ran back toward the stall where Shelba had been packing.

Twenty minutes later on the overpass leading out of Old Town Paducah, George inched the van forward another 10 inches, again touching the bumper of the car in front of him. The pickup on his right tried to push into the non-existent space between the two vehicles, and George did not intend to let it. George's vehicle could not move in any direction; the jam had reached the point of maximum compaction.

"What the hell are those bastards doing? It looks like people are stopped on top of the overpass."

Shelba looked out the rear view mirror on her side. "George, I can see water coming up between the cars behind us. Maybe we should just get out and walk ahead."

George looked at the scene behind him. "Oh God, you're right. The water's rising faster than we're moving. Why don't those people move out ahead?"

"George, it looks like there's some kind of flood on the other side of this overpass. Maybe there's no place for anyone to go. Let's get out and walk ahead, please."

"If they'd just move ahead a little, we could get ahead of the water. What's wrong?"

Shelba could take the strain no longer. "George, I'm getting out and walking. This van's not going anywhere, but the water's coming up really fast. There are other people walking, so I'm walking, too. Are you coming?" She opened the door and stepped out onto the pavement. The water had already crept to within 50 feet of their van, and when it's level rose another two feet, it would cover their wheels.

"What about our exhibits?"

"Now is not the time to worry about what we have in the back of the van. Now is the time we worry about our lives. Are you coming?"

George turned off the engine. Opening the driver's door he stepped out and slammed the door shut. "Dammit, why don't they have some decent roads out of this city? Why don't they plan for something like this? Where have they been hiding their heads?"

He joined the growing throng climbing over the packed vehicles and walking up the incline to the top of the overpass. None of them considered that even the elevated overpass might not be sufficiently high to stay above the floodwaters that would soon engulf the city.

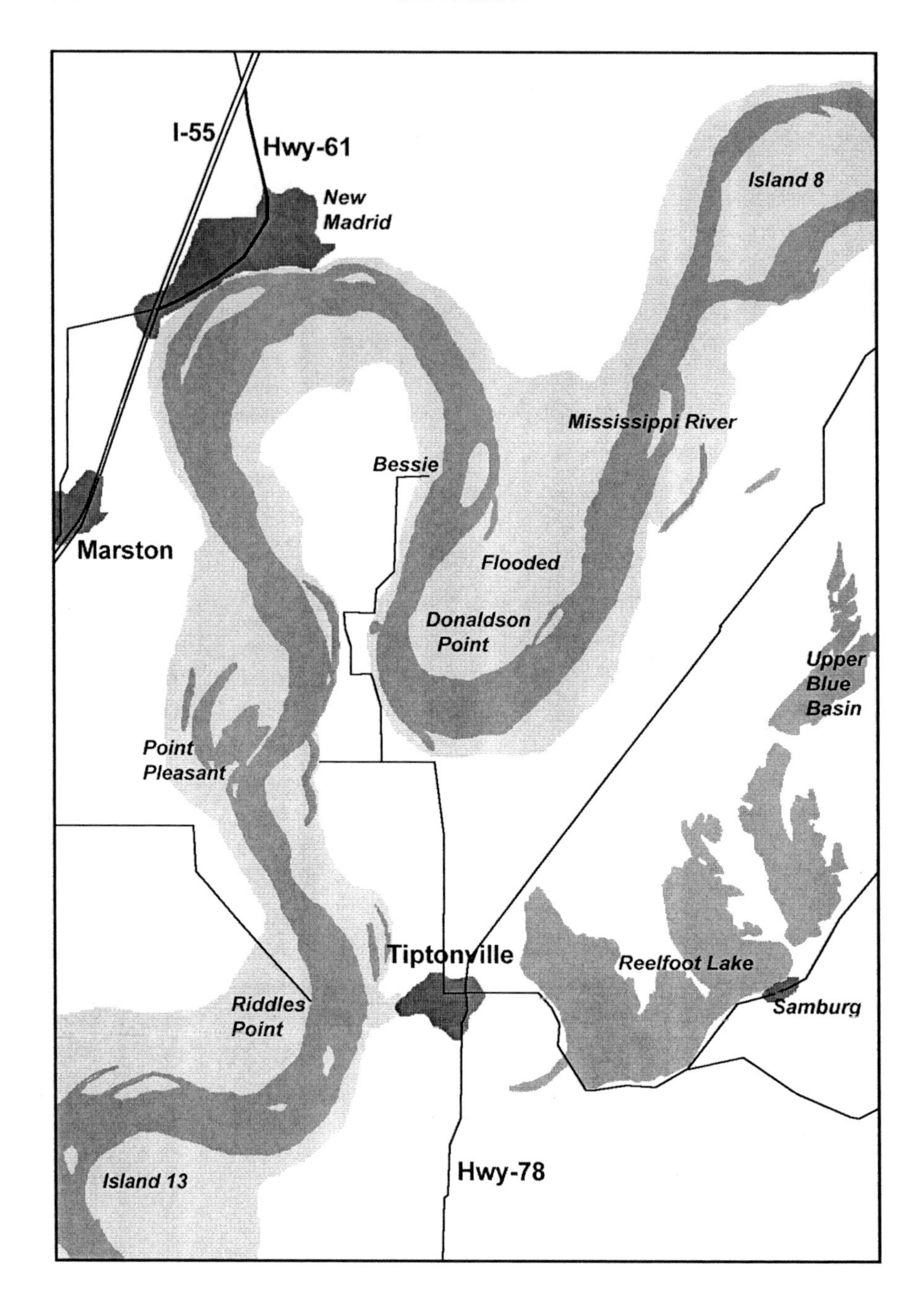

New Madrid Loop and Bessie Landing

— 8 —
TIPTONVILLE

"Cap'n, according to that book about the New Madrid Fault, they had lots of aftershocks from the earthquakes in 1811 and 1812, and some of them were really big. Do you think we'll have any big aftershocks?" Jeff stood behind Buddy Joe in the wheel house as they monitored the radio chatter.

"That's a major worry, Jeff. Yes, there will be aftershocks. The question is how big and for how long. They say at least two of the biggest earthquakes they had in 1811 were aftershocks of the first one, and the shaking didn't die down enough for people to live normal lives for five years."

At 10:45, one hour and 41 minutes after the initial temblor, the New Madrid Seismic Zone experienced its next major seismic event, demonstrating its propensity to produce a series of large temblors.

The new fracture began 14.3 kilometers beneath the surface with an epicenter just two miles southwest of Tiptonville, Tennessee, only a mile from where the initial fracture had ended. With a magnitude of 7.4, this fracture occurred in the thrust portion of the fault zone, rupturing another 500 square miles along a crack last displaced 1,140 years ago.

A block of the earth's crust northeast of Tiptonville pushed to the southwest under another block along a fracture plane that dipped 30 degrees from the horizontal toward the southwest. The old crack extended over a trapezoidal area with corners at Obion and Dyersburg, Tennessee, and New Madrid and Portageville, Missouri. Maximum displacement along the plane of the fracture near the focus amounted to 11 feet. The maximum vertical lift of four feet occurred near Ridgely, Tennessee. The fracturing lasted for 15 seconds before energy stopped flowing from this latest break in the earth's crust.

Cities and towns to the north of the epicenter felt the brunt of the earthquake waves. The shaking intensity in Tiptonville, Bessie, New Madrid, and Marston ensured that nothing remained standing, not even New Madrid's famous telephone building that had been built within a protective cage. The remaining bridges crumbled. Broad fissures criss-crossed the ground and liquefaction filled the fields with sand boils. Reelfoot Lake sank three feet.

Sikeston to the north and areas to the south of Caruthersville and Dyersburg felt accelerations exceeding 60 percent of gravity. Most masonry and frame buildings, including any historical buildings still standing, crumbled around their broken foundations. Fissures appeared in the ground and ground water spewed forth. More landslides leveled any higher ground.

Cape Girardeau and Paducah felt another round of shaking with accelerations half that of gravity. All buildings suffered some damage and any poorly built structures left standing were leveled. The shaking knocked frame houses off their foundations and conspicuous cracks crossed the fields and lawns.

The urban centers of Evansville and St. Louis and other cities a similar distance away experienced another 60 seconds of shaking at least as strong as that in the initial shock, hurrying along their urban renewal projects.

Once again the waters of the rivers frothed and danced. Once again the shaking loosened the soils comprising the banks, levees, and bluffs along the rivers and hurried them along on their march toward the river bottom.

Steve had traveled half a mile when the levee began to shake again. "Hold on, we're having a heavy aftershock. That is what I was afraid of." He shoved the accelerator to the floor.

The thrust fracture of the Tiptonville aftershock, its epicenter less than 10 miles away, ran 16 kilometers deep beneath Bessie's Neck. Acceleration forces on the levee exceeded half the force of gravity.

The levee sagged unevenly, more and more. Steve saw rivulets running over the road atop the dike ahead. He had no choice. Pushing even harder on the accelerator he steered the drunken pickup along the ruts on top of the levee, through streamlets that grew to streams before his eyes.

Steve bounced on the old vinyl seat and yelled. "Juliana, hang onto something. We've got to get off this part of the levee as soon as we can. It's dropping too far too fast." He grasped the steering wheel; Juliana just bounced with no place to attach her hands.

"Daddy. Up ahead." Juliana screamed a warning. "The river has cut through. It's too deep."

"Hang on." Steve pressed even harder on the accelerator, demanding more speed from the old pickup. The truck fishtailed on the dirt as it sped toward the growing swath of flowing water. "And pray," he yelled.

A rolling seismic wave swept across the land from the south and lifted the pickup as it careened toward the cut in the levee. The truck leapt halfway across the flow before it dropped into the water, sand, and mud. It caused a huge splash of water to each side of the truck and to its front. Muddy water covered the windshield an inch deep, and Steve had no idea if the truck still pointed in the right direction.

Steve stomped the accelerator pedal harder into the floorboard and wiggled the steering wheel, searching for some feel of solid land. He kept the wheels spinning and the truck bounced enough to find purchase on the fast moving soil. A sudden rebound lifted Steve off the seat as the truck splashed its way up the opposite bank onto dry land beyond. The truck careened toward the left side of the dry road ahead, and Steve jerked the steering wheel hard to the right, bringing the truck back into the ruts.

Looking in the rearview mirror he could see that the water had torn away more and more of the underpinnings of the road as it spilled down the landward side of the levee. He could see waterfalls developing on the levee, cutting deeper and deeper into the roadbed.

The shaking quieted, and then stopped. Steve slowed. The washing of waves from the river onto the road subsided, and the chop on the surface of the river smoothed. The water came within a foot of the top of the levee along which he now drove, but it had not come over the top, at least not yet.

"We made it. God, that was close." Looking over at his daughter he saw her crying and biting her lower lip. "Honey, we made it. We're okay. We're okay."

Steve stopped the truck where the levee road reconnected with the asphalt highway to Bessie. The road had climbed a little higher and farther away from the river waters. The levee on which the truck rested appeared stable. He killed the engine and turned to look out the rear window as he reached across to comfort his daughter.

Left to its own, the crevasse would take several hours before the cuts in the levee would grow to the point of total failure, though the occasional small aftershocks hurried the process along. The insufficient flow would not yet erode away all the sands that formed the levee, but the river water had found the path it needed to cross the neck and rejoin the river half a mile away. The inevitable required only a matter of time.

Charlie Green steered his 39-foot cabin cruiser at quarter speed up the river toward Osceola through the Driver Cutoff channel, approaching the sandbar at Plum Point Reach, a mile and a half above Fort Pillow. His back still ached from the pounding his boat had received an hour and a half earlier.

He and Sylvie had watched with amazement as waves and choppy waters surrounded their boat when they lay anchored off Fulton. The river churned as if a gigantic frothing monster swam just beneath the surface. No black clouds filled the sky, no breath of wind stirred the air, and yet the waters of the river thrashed as if caught in a gale. Only when Charlie had heard frantic radio calls from several marinas along the shore begging for help did he realize the full extent of the disaster.

Now the river came to life again. "Sylvie, there's another earthquake that's causing this rough water. It must be a big aftershock. There's no way to tell if it's bigger than the last time or not, but it is big."

"Charlie," his petite wife called from the stern. "Back there where the girls are trapped. It looks like the bluffs are sliding down some more. I see dust and spray around the bottom." Sylvie again scanned the bluffs through her binoculars. "Yes. The entire bluff is falling into the river. There's a whole line of trees dropping. It's just black and red dirt behind it. Charlie, it's huge." The alarm in her voice startled her husband.

Looking back, he could see the slide without the aid of the binoculars. "That's going to create one hell of a big wave," he said. "This chop will feel smooth by comparison. Come up here beside me."

"Oh, those poor little girls. They'll drown. Charlie, we've got to go back and try to save them. Maybe we can take them back to Memphis. Can't we go back? I'm afraid." Sylvie reached back and touched her husband's arm.

After a minute the shaking slowed and then ended. "I suppose you're right. We need to get off this river, and Memphis may be a better place than Osceola if this shaking keeps up. We need to find some place that's safe. And we can look for the girls along the way, but Sylvie, I don't hold out much hope for them. That was a really big slide."

Charlie spun the wheel to starboard and throttling up to 10 knots and headed the boat back down the river. He peered to the east and south. "I'm not going to go too fast. We still have to cross the wave that's coming from the bluffs."

He squinted, trying to make the distant image clearer. "You know, even from here I can see where the swell is heading upriver at us. That has to be some big wave to see it this far away."

He watched a moment longer and then commanded. "Sylvie, go below and get our life jackets, really quick. We may need them. Hurry it up." He watched with growing alarm as the wave crested over a bar near the shore a mile away. The crest grew to 12 feet high as it crashed into the trees beside the revetment, like a tsunami coming out of the ocean.

"In fresh water they call a wave like that a seiche," he called over his shoulder. "God, I can't get caught over shallow water by something like that. It'll swamp us for sure." Checking the depth meter, he saw the boat moved in only 10 feet of water, definitely the wrong place to be. He swung the boat hard to port and gunned the engines, racing for the buoys marking the main shipping channel. It became a sprint between his boat and the approaching wave.

"What's wrong, Charlie?" Sylvie climbed into the pilothouse with the life preservers. "Why did we speed up?"

Charlie gripped the wheel, estimating what his position would be when the wave struck the boat. "I'm running for deep water. I've got to get there before that wave hits us." He explained his plans to Sylvie. "Get hold of something. Just before the wave peaks, I'm turning hard to starboard and head directly into the crest."

"But you're going right where the wave will be. Why not go upriver?" Sylvie became more and more alarmed, almost hysterical.

"We can't outrun that wave. It's coming too fast. Put that life preserver on, now," Charlie ordered as he slipped his arms into the straps of his life preserver and pulled the tightening cord. He could see the swell on the surface approaching the boat, closer and closer. "Hang on," he yelled as he swung the boat back toward the approaching wave and cut to half throttle.

The water rose in front of them. The boat had not quite cleared the edge of the bar. The wave grew and grew, and Charlie pushed the throttle forward to challenge the water.

The large boat rose into the wave, and the wave crested just as the boat reached the top, covering the bow of the boat for an instant. The hull hesitated, hung suspended at the crest, then it slammed forward and with a slap headed down the back slope of the wave, once again racing toward the deeper waters to the south. Behind them, they could hear the wave break with a roar and crash down as it rushed over the bar and on up the river across the islands of Plum Point Reach.

Charlie cut back on the throttle as they moved into the deeper water. He grabbed his wife in his arms and they turned to watch the wave continue up the river. Along the shores and on the islands it grew into a monster tidal wave, sweeping the levees and flooding the low-lying lands. "My God, what will happen when that wave reaches Osceola?" Charlie had never imagined there could be anything on the river like what they had just seen.

The Tiptonville fracture ruptured the rock matrix to within six and a half miles of Caruthersville. Again, like the shock wave that precedes an airplane breaking the sound barrier, much of the S-wave energy coursed through the town in an initial 13-second burst, shaking the town like a ferocious dog.

Those who had chosen to stay in the town saw the levee that was supposed to protect the streets from the river sink further. Water flowed over the sandbags and spread through the streets and into the lawns of the houses. The center of town sagged another two feet below the level of the river.

As the shaking continued the levee to the north of town, which now held over 500 refugees, settled faster than anyone could imagine. Within nine seconds it disappeared beneath the river waters, leaving the lost souls to swim for their lives in the muddy flow that rushed across their former refuge.

Minutes after the shaking ended, the tallest structure that remained in Caruthersville was the southwest corner of the courthouse. Loretta pulled the old woman by her shoulders up its concrete steps.

"Thank you, dearie. I still don't understand why the 911 people never came. I was beginning to worry some. I don't know what I would've done if you hadn't come along. That water was halfway covering my legs."

Loretta said nothing but kept dragging the woman higher and higher away from the rising waters, stopping when she backed into resistance. She stood and turned to look up into the eyes of one of the deputies. "Sorry, Loretta, we ain't got no more room inside. You'll have to leave the old woman out here."

"She has a problem with her leg. I need to get her inside and find some medical help. Please move out of the way." Loretta started to reach down to continue her mission.

"I said there ain't no more room. I kin find a place for you upstairs, but the old woman stays here. You said she cain't stand, so she couldn't make it up the stairs anyhow." The deputy spread his booted feet apart to impose a larger barrier.

"Now, I know who you are, Loretta, 'cause we got a book on you, and I said I kin find a place for you." He licked his lips. "You want to go on up? I kin meet you up there in about 15 minutes. You do understand, don't you now?"

Loretta felt shock and dismay. She had been a working girl long enough to recognize the implication of what the deputy suggested. "Not now, I'll just wait here." She sat down on the steps next to the old woman and took the old woman's crooked hand in her own.

Buddy Joe leaned back on the skipper's stool, cuddling a cup of lukewarm coffee in his hands and waiting for more information as the radio scanned the marine frequencies for activity.

The monstrous aftershock had done little to the boats that they had not experienced before. The hardened refugees who had transferred from the overflowing *Bella Queen* and sat scattered across the barges had ridden the bucking ride with a remarkable lack of panic.

Snitches of radio traffic told him damage had reached north past St. Louis, east on the Ohio into Cincinnati, west on the Missouri to Kansas City, and as far south as New Orleans. There was even some word that the Achafalaya had captured the Mississippi. But still no one had taken charge to tell them what to do or where to go to find safety.

He thought about his wife Helen, wondering if she had been hurt, wondering what her sister's home in Millington down by Memphis must look like now. Being inland there would be no fear of flooding, but fires and falling structures like those around Caruthersville must have been repeated throughout the whole area. Would she blame him for all this, too?

The abrupt sound of the radio brought Buddy Joe back to the present. "Calling anyone still on the air. Anyone read me?"

Buddy Joe picked up the microphone to respond. "This is Captain Buddy Joe Simpson of the towboat *Lady Bird Jamison*. I'm just above the Boothspoint Bridge with the *Bella Queen* excursion boat. Who's this? Come back."

"Captain Simpson, my name is Charlie Green. My wife and I are on our 39-foot motor yacht, the *Amanda Blair* out of New Orleans, motoring just south of Osceola. There's just been a monstrous landslide down at Fort Pillow that sent a huge wave upriver. It appears to have taken out the Osceola harbor, probably the town, too. I've been trying to contact someone to report it, but all radio communications down here seem to be off the air. There's just this God-awful static. Over."

Buddy Joe adjusted the volume and thumbed the microphone, "I don't know where the static's coming from, but it's been that way ever since the shaking started. What's this about a landslide and Osceola?"

"At least half a mile swath of the Chickasaw Bluffs caved in and fell off into the river with this last big aftershock. We were two miles upstream at the time. When the wave passed us in the middle of the river it was at least 12 feet high. Along the banks it appeared to be nearly 20 feet. It headed directly into the Osceola harbor. I'm near the south entrance now and there appears to be substantial damage along both shores."

Buddy Joe asked, "Do you see any boats up in the harbor? My sister ship, the *Lady Jane Wilson*, made a delivery there earlier this morning."

"No sir, there are no boats moving on the water. It appears some larger boats washed ashore up near the town, but that's about a mile away. I can't identify any of them. At this point we're looking for victims, but holding in the middle of the channel and staying away from the shores. I was wondering about the status elsewhere along the river."

Buddy Joe shuddered, thinking of the fate of Paul Taylor and his crew.

"Charlie, Caruthersville is pretty well wiped out. The skyline's gone and there's lots of smoke. All the grain elevators either exploded or crumbled. I see no indication that anyone's fighting the fires. Most of the levees are down with significant flooding. I'm hoping for someone to do a check of the I-255 Bridge for us so an excursion boat and my tow can proceed downstream. After the flood surge comes through I expect we'll head downriver to Memphis. Do you have any reports from below you?"

"Not really, but what's this about a flood surge?"

Buddy Joe related what he knew about the breaking of the locks at Kentucky Dam and the schedule for changes in the water levels along the river. "Charlie, keep monitoring this frequency in case we need to get in touch. You can also hail my sister ship." Buddy Joe repeated the hailing frequency for the towboat fleet. "I'll keep scanning for other radio traffic. Boats on the river may be the only ones who can communicate. There has been some UHF and VHF ham radio traffic, but only battery-operated units are broadcasting. All other land-based radio seems to be down."

"I'll keep in touch, Captain Simpson. I'll report back to you on status as we move on downriver. Maybe we'll meet up in Memphis."

"I look forward to that. *Lady Bird Jamison* clear." He pushed the button to continue the scan.

The old woman leaned over and whispered to Loretta, "You should've taken him up on his offer, dearie. It's your only chance to make it through all this."

Loretta looked the old woman in the eye, wondering from where the old crone's advice came.

The old woman continued, "I understand how much trouble we're in. The river's going to keep rising and this town will be flooded even deeper. You're young. You'll have to give him what he wants, and if you do then he'll take care of you."

Loretta's vision of the old woman clouded.

"Dearie, I been there. I understand. I had to work the streets, too. You got to take the chance when it comes." She squeezed Loretta's arm with her gnarled, bony hands. "Thank you for coming back for me. That was a right kindly thing to do. Now go."

Loretta's head and shoulders drooped. They ached, her eyes stung, but she stood and turned back to the deputy. "Which room?" She felt tired and dull, like she often felt when she worked. Only this time the price was her life.

— 9 —
BOLD MANEUVERS

Minutes after the first earthquake the air conditioning had been turned off in the *Bella Queen* lounge and the doors flung open. The hot, humid air made no difference; people filled every corner of the room. Lynn leaned against Ron's knee on the floor in front of the TV.

The TV signal came from a satellite dish atop the pilothouse. The clock behind the bar showed three o'clock as the Purser turned on the receiver and TV; the picture brightened to show a scene from an airborne camera as the feminine voice of the News World Satellite TV anchor explained.

"... waters of the Mississippi River flowing from the many breaks in the levee system across the farmlands north of Cairo, Illinois, all the way south to New Orleans. In southeast Missouri much of the water is flowing out toward the St. Francis River drainage basin, following the drainage ditches that trace through the flatlands such as these we see to the south of Caruthersville, Missouri. These pictures are provided by our affiliate in Little Rock.

"Much more flooding is predicted as the flood surge from the broken locks at the Kentucky Dam is reaching this area and raising the level of the Mississippi in northern Arkansas by more than five feet. Before the earthquake, the river ran six feet above the flood stage. Aerial photography indicates this morning's shaking caused most levees to collapse entirely or to drop to near the water level."

The scene changed to an urban area along the river and again showed breaks in the levees at several points. "Pictures of the flooding in and around New Orleans come from our Baton Rouge affiliate. The Army Corps of Engineers expressed doubts that any of these levees can be repaired before the flood surge reaches the lower Louisiana delta three days from now.

"The Army Corps said the loss of the Old River Control Structure above Baton Rouge has released over two thirds of the Mississippi floodwaters into the Achafalaya, easing the pressure on New Orleans. However, Morgan City along with the other small towns downstream on the Achafalaya River are expected to be total losses.

"State and Federal authorities have ordered mass evacuations from the southern half of Louisiana in hopes of minimizing further loss of life, but most of the escape routes have already flooded. Water pouring through the Achafalaya basin has already destroyed the I-10 causeway, cutting off escape to the west.

"Changing scenes, footage we showed earlier, provided by our affiliate in Indianapolis, shows the waters gushing through the broken locks at the Kentucky Dam just east of Paducah. Officials now estimate the water is flowing at a rate of 55,000,000 gallons per second. Water is already 15 feet deep in Paducah and is starting to rise in Cairo. This break will cause major flooding downstream on the Ohio and Mississippi Rivers for the next two weeks.

"The Army Corps engineers estimate it will take at least 10 days for the flow to drop to under a million gallons per second. Officials declined to answer questions about ways to stem the flow before that time.

"In Paducah, authorities are desperately trying to organize rescue boats to save the thousands trapped in buildings and on overpasses in the flooded areas. They report that the water rose so quickly that few had a chance to escape."

When the TV screen displayed anchor Carolyn Phelps sitting at the NWS News Desk, Lynn whispered to Ron, "She looks like she's in shock."

Carolyn stared into the darkened studio beyond the cameras in response to some unheard command. She nodded her head and turning to the cameras attempted to put a smile on her face. "Excuse me. We just felt another bit of shaking. I am still tense from this morning when Atlanta was rocked by the earthquake. Please pardon me, but everyone here is still jittery."

New footage appeared on the screen and she continued, "Our Nashville affiliate's cameras show a huge fire raging out of control just south of Memphis along the President's Island shipping channel. This is the site of a large number of refineries and chemical plants. Toxic smoke from this monster blaze continues to drift to the northeast across the eastern half of city and on toward Jackson.

"Views of Memphis show major destruction of commercial buildings along its bluffs. There appears to be widespread damage elsewhere throughout the city. There has been minimal communication with authorities in Memphis because all communications links to the city have been destroyed. Only a few radio facilities with backup generators have been able to come back online, and those are fully devoted to rescue efforts.

"In St. Louis major damage was recorded when several parking structures pancaked near the Arch and when the old Union Station collapsed atop the Saturday morning tourist crowd. There are reports of over 500 casualties in Union Station alone. Amazingly, the Arch remains standing though we understand there was a major loss of life in the museum beneath the structure.

"East and south of St. Louis, near the town of Marion, Illinois, there has been a major collapse of an old underground coal mine. Smoke from fire burning in the pit covers the area. The extent of casualties at that site is not known, but over half the town was consumed by the collapse.

"In Washington, the first figures have been released from the Federal Emergency Management Agency regarding estimated casualties and damage. They now believe as many as 8,000 people could have been killed and 50,000 injured throughout the region affected by the earthquake and floods, but they do not have an actual count as yet. They estimate there could be over a million people left homeless. Damage is expected to run over 50 billion dollars.

"Others have suggested that these figures are far too low and that deaths and injuries will number in the hundreds of thousands and damage will be 10 to 20 times the initial estimate.

"Cal Tech in Pasadena, California, reports they have recorded four aftershocks measuring above magnitude 7.0 since the initial shock of 7.9 at 9:34 Central Daylight Time this morning, the largest the 7.4 magnitude event at 10:45. Some seismologists have expressed grave concern because of the New Madrid's history of producing a series of several very large earthquakes. After the first earthquake in 1811, four more giant earthquakes measuring above magnitude 8.0 occurred within the next two months.

"I understand we are now ready to show some new video taken from a small airplane flying over the Memphis area. Be aware these scenes are graphic and disturbing in nature." The TV screen filled with a mosaic that in time people in the lounge realized focused almost directly down on the Memphis area. A large black smudge of smoke ran from the lower left of the screen across the middle and out the top.

"The camera scene is moving to the north as the airplane flies in from Mississippi. The smoke you see comes from the huge industrial fire covering the majority of President's Island and the shipping channel. Smoke from that fire is being blown across parts of Memphis and Germantown."

The camera slowly zoomed in on the downtown area. Streets appeared bordered by blurry buildings, and then it became apparent that the camera revealed the remains of buildings smeared across the landscape. Smoke billowed from many structures scattered through the city.

"There appears to be very little fire-fighting equipment at any of the fires you see in the city, probably because most or all equipment was destroyed in the earthquake."

The camera zoomed even closer, showing people working amongst the ruins. On one street, bodies of men, women, and children lay side by side in a long row, as if asleep. Lynn heard mutterings and gasps from the audience behind her when they realized that they were looking at the dead. She too had trouble breathing. Her breaths came in gasps from the emotional clutching of her chest and throat.

"Finally, we have these pictures of some of the high-rise buildings in the center of Memphis. As you can see, several of these have fallen, and others are still in danger of collapsing. People have left the heart of the city in droves and are seeking shelter in any open space they can find."

Carolyn received a handwritten note on camera. "And this just in. The President of the United States has ordered that Armed Forces and National Guard Units across the country be mobilized immediately to aid in rescue efforts in the stricken areas. He declared the central United States a Disaster Area and imposed Martial Law in areas of the seven states most severely affected by the earthquake: Arkansas, Tennessee, Mississippi, Louisiana, Missouri, Kentucky, and Illinois."

Carolyn again painted her professional face with a smile as she looked up into the camera. "And now the international news. Offers of help are pouring in from countries around the world. The State Department has been given the responsibility for organizing these and …"

The purser looked at his watch. "I'm sorry but the 10 minutes I was allowed have stretched into 12, and Captain Ruggs was explicit about not using too much power." He reached over to turn off the receiver and TV.

A few cries of dismay sounded, but when Lynn looked back at the audience, she saw that most of them sat in stunned silence, their faces streaked by tears.

"We'll take another look in a couple of hours. In the meantime, there are many around us who need help," the purser said. "Please go out and breathe some fresh air and find ways to aid or comfort your fellow passengers. And pray for those you just saw on the TV. As you can see, we are blessed by where we are."

"Dammit, Barney, you're a fool to move your boat now. We can see there's been damage to the Boothspoint Bridge, but we don't know how much. Stay away from that structure until we can get a better idea of what's wrong." Buddy Joe tried his best to talk Barney Ruggs out of his idea to head downstream, but Barney's mind remained set.

After taking on survivors from Caruthersville, the *Bella Queen* followed by the *Lady Bird Jamison* with its tow had moved to within half a mile of the large bridge crossing the Mississippi River between Missouri and Tennessee. From that distance they could see damage to the causeways, but the shipping channel appeared to be open.

"Buddy Joe, I've got a ship full of passengers that want me to do something. We had satellite TV going in the lounge and they've been watching the news. The President has declared Martial Law. The news reports say much of the Mississippi valley has been destroyed, and they showed some pictures of the break in the Kentucky Dam."

Buddy Joe interrupted. "That just proves my point. You'll be jumping out of the frying pan into the fire. Let's just hold here until we have a better handle on where to go."

Barney argued back, "With that surge coming down the river at us, it'll push us into the bridge or maybe even collapse it in front of us. And if Bessie goes like you think it will, we'd be stuck here with nowhere to go. I'd just as soon be farther below than here. We can see the causeways are down, but the main truss of the bridge is okay."

"It's your decision, Barney, but I still say you should wait until somebody comes along in a small boat to check out the passage under the bridge." He waited, as if hoping for a response. "But if you're serious, I'll hold up here and be ready to help, if I can, as you go down."

"I'm going." Barney hunched his shoulders, making the final decision to exercise his command as the Captain of the *Bella Queen*.

"Sylvie, I know Memphis must have suffered major damage from the earthquake, but I agree we should head back to the Wolf River Marina for now. Osceola is a total mess from that wave and I haven't seen any refugees coming out of the harbor. I just don't want to go up into that mess."

Charlie steered the boat into the middle of the river, letting the current carry it downstream. Unsure of what he might find, he watched the depth meter and the river ahead with care.

He soon noticed that the current had slowed. He had enjoyed fishing this part of the river the last few days and had come to know it well. Now it did not feel right, not as he remembered. He scanned the water toward Driver Cutoff and could see the river running high against the bank. "Sylvie, that slide must have really changed the river." He motored with caution toward the west bank and the current began to flow faster and faster.

As the boat rounded the curve, it took several minutes before he recognized the head of a new huge rapids in the river along the west bank. "Sylvie, that slide dammed the whole east side of the river. It's pushing all the water through a chute on the right." Giving the boat full throttle, he turned the tiller to head back upriver. The current sucked at the boat and pulled it deeper and deeper into the channel. Charlie gunned the engine and steered toward the slower water to port, closer to the east bank.

Several minutes passed before the boat rounded the head of the rapids. "God, Sylvie, the depth finder shows it's only 12 feet in through here. This was 60-foot water this morning. That slide almost filled the entire river." He motored along the top edge of the underwater dam, tracing its contour with the depth finder.

Charlie pointed ahead. "Keep an eye out for any kind of debris or brush in the river. It may be just below the surface."

"Charlie, look over there, in those trees close to the bank. There's someone in the treetops." Sylvie grabbed the binoculars and examined the branches and leaves protruding from the water. "Charlie, it's those two girls we saw this morning up on the bluff. What are they doing down here in the trees? You don't think they rode the landslide all the way down, do you?"

"I sure don't know how else they'd get into that predicament."

"We've got to save them. Charlie, we must."

"I know." Charlie understood what fate prescribed. "I'm already heading over to them. You get the boat hook. I don't know how deep it is next to the trees or what kind of trash we'll run into." He idled the engine to drift the boat toward the two waving girls, now close enough to hear their calls for help and screams of joy.

Sylvie ran up the side deck to stand on the bow with the boat hook in hand as Charlie approached the tree. "Charlie, run the anchor out about three feet so it hangs close to the water."

Charlie pushed the RAISE button forward to pay out the line.

Sylvie caught a tree branch and pulled the boat closer. "Amanda, when I get close enough, grab the anchor rope. You can step on the anchor and crawl up onto the boat. Blair, you just wait your turn. We'll get your sister on board then pick you up."

Danielle looked through the branches to Samantha and asked, "Who's she talking to?"

"I don't know, but they're here to save us. Don't argue."

Sylvie tugged on a limb to pull the boat close to Samantha who climbed onto the anchor as instructed. From there she climbed over the transom and into the boat.

"Okay, Blair, now it's your turn." Using the boat hook Sylvie pulled the boat over to the next branch so Danielle could reach the anchor rope.

Danielle screamed and jerked back. Charlie saw her almost fall from the tree. "A snake," the girl screamed. "It bit me." The serpent, nestled in a higher branch of the tree, sank its teeth into the girl's forearm. She shook her arm but the snake remained attached, its teeth anchored into her skin and flesh.

Sylvie dropped to her knees and knelt over the edge of the boat. She grabbed Danielle's other arm and hefted her 70-pound body onto the deck. Sylvie turned and grabbed the snake at the back of its head, squeezing and opening its jaws even wider until she could disengage the short teeth from the girl's arm. Danielle screamed and wrapped her hand around the bleeding wound.

"Blair, I know it hurts, but it's only a water snake. It's not a cottonmouth." Grabbing the snake's tail, she slammed the snake against the deck then flung it out into the water where it surfaced and swam back toward the trees.

Turning to the screaming girl, she took the arm and inspected it. "There, there, it's okay. I'll put some antiseptic on it and bandage it up. You'll be okay. Just be calm." She hugged the hysterical girl. After a long minute the screams subsided into deep sobs and then uneven breaths.

The older girl walked across the forward deck to the pilothouse and spoke to Charlie through the side window. "Thank you for saving us. We were beginning to get worried. My name's Samantha McCutchen and that's my sister, Danielle. We live on the other side of Fulton."

Sylvie led Danielle over to Samantha and brought the two girls into her hug. "Oh, it's so good to have you girls back. You look so fine and now you're safe. I was so worried about you, Amanda. I didn't want anything bad to happen to either you or Blair. Now you're okay." She hugged them tighter and tighter and cried.

Charlie could see the girls growing uncomfortable. He shook his head as he backed the boat away from the brush into deeper water. He explained to the girls, "Please forgive her. It's this earthquake. Sylvie just can't remember your names."

Barney rang for half-speed-ahead to the engine room and began to execute the tight 180-degree turn to starboard to bring the boat in line with the bridge. Dropping back to quarter-speed he steered the *Bella Queen* on its slow journey down the river toward the I-255 bridge. As the riverboat came closer to the bridge, he eyed the truss arches with suspicion. "Mate, keep your eye peeled on that bridge and let me know if anything's about to fall on us."

Passengers and crew stood forward on all decks of the boat, watching the mighty bridge above them and the ravaged land around them. Few watched the muddy river below.

When Barney sounded the boat horn he saw an old black man standing in a small boat next to the south pier of the bridge. The man waved his arms back and forth, as if signaling the boat to stop. Passengers on the boat waved back in a joyous mood and holiday spirit.

The man yelled something, but Barney could not hear him. The riverboat continued heading down the middle of the shipping channel toward the bridge.

As the boat entered the shadow of the bridge, Barney and some of the passengers became aware of the missing portion of the roadbed on the west side of the bridge. Barney swore under his breath. "My God, it looks like half the roadbed must have fallen into the water."

Most of the passengers continued to look at the gaping hole above them, but some, including Barney, searched the river ahead. Scanning the water he spied a vague dimple in the surface showing where the fallen concrete had become what in the old days would be called a sawyer. Like the roots of a sunken tree, the roadbed stood on its side with a ragged corner reaching up into the heart of the shipping channel, four feet beneath the surface.

The riverboat's course tracked straight toward the snag. No one suspected there would be any danger in the river at that point.

Barney slammed the General Quarters button to sound the sirens and steered hard to starboard. The bow of the boat missed the piece of roadbed by a few feet, but the current of the river carried the stern sideways over the concrete and steel snag just beneath the surface of the water. Five seconds later he heard a ripping squeal as the bridge span caught the steering and propellers of the riverboat and bent them across its bottom.

The sudden stop threw passengers forward, and some fell to the decks below. Two men on the second deck, who had leaned far over the rail to watch the water, fell with a splash into the muddy river. One of them was first mate Ralph's son, Ricky. The rapid current swept them downstream, bobbing in the water and shouting for help. They disappeared around the bend flailing in the water trying to swim for the safety of the levee shore just a few yards away.

Barney looked back and forth across the river, trying to gauge what had happened to his ship. "Dead in the water. Hung on a snag. Damn." The boat swung back to port, pivoting on the slab, and weaved back and forth in the current under the bridge. "Engine room. Report. What's your status? Check for leaks and damage."

After 30 seconds Barney received a response from below. "Captain, we don't see any leaks around the engine room or steerage, but the rudders and props appear to be bound on something. It's like we run aground. What happened?"

"We did. We're caught and hung solid. The roadbed from the bridge above us fell into the water and we're locked solid on it. To say your Captain screwed up would be an understatement."

— 10 —
Preparations

One hundred and twelve miles upriver, Paula leaned against the steel bulkhead of the tiny tool room. She opened her blouse and rubbed the nipple of her swollen breast across the baby's mouth. The eager infant responded and began to suckle milk from its mother. "Oh, you're so beautiful. My baby girl is just like a little doll." Paula cooed to the child and rocked the naked infant in her arms.

Freddy squatted on the deck outside the compartment and watched in awe. He reached in and rubbed the baby's brown arm with the back of his finger, bonding with the new family. A clap of distant thunder reminded him of the mid-afternoon thunderstorm coming their way.

He found an old bucket and some line and lifted a supply of Ohio River water to the deck. As Paula washed the new baby in the water, the pair watched more land around Fort Defiance sag beneath the river waters as aftershocks kept the soil the consistency of sloppy mud. A brief rainstorm pummeled their small car still sitting in the parking lot, now half submerged in the encroaching waters.

Paula peeked out from the shelter of their haven. "Freddy, the river bottom must be sinking. See how the rope holding the barge is sloping down more and more. Do you think it'll pull the barge under?"

"I'll loosen it, then it'll be okay." He worked on the line, pulling it free of the cleat to pay out more slack.

"Freddy, our car, it's sinking under water awfully fast, too."

Freddy studied the car, other barges, the trees and bridge structures lining the river. "Paula, something's wrong. All the barges are moving around." The line became taut in his hands. Paula watched him lean back to hold tight, but slick from the rain, the rope slipped through his hands so fast the end of the line whipped through his grasp before he could react, vanishing into the water. "Paula, I lost the line. We're drifting."

She clambered out of the little shed holding the baby. "Freddy, are we okay?"

He waited a moment before replying. "Something's really wrong." The barge drifted past the trees marking the point of land of Fort Defiance into the Mississippi channel. "Look, this barge is swinging around and we're being pushed upriver into the Mississippi. And the water is already halfway up the observation deck on Fort Defiance. It's flowing overland into the Mississippi. It's the Ohio River that's gone wrong. It's rising, really, really fast. So fast it's making the Mississippi flow backwards."

The pair rode in silence and watched as their barge floated two miles upstream in the muddy Mississippi River and the Ohio back-filled the larger river with water flowing from the Kentucky Dam locks.

Over an hour passed before Paula remarked, "Freddy, look, we've started moving back downstream. I can tell from the trees along the bank." Later, as they passed beneath the Highway 60 bridge near the water-covered junction of the two rivers, she spoke again. "I guess I'm getting my wish."

"What's that?"

"To go away from here." Paula hugged her newborn to her breast and she reached out to cover Freddy's hand. "I don't know if I should laugh or cry."

On the side of the interstate overpass west of Dyersburg a small crowd had gathered to stare across the water-covered flood plains of the Mississippi. Most sat on the ground, resigned to the next aftershock that would rattle the ground beneath them. They all displayed symptoms of mild shock.

Chris Nelson and Alex Smyth sat in the shade a few yards away from the flood. Limping up behind them, Tina Washington could see the blood that stained Chris's blond hair. His head must be hard as a rock, she thought. He's so lucky that overpass didn't crush him inside his pickup. He came out of it all with only a minor injury. Memories of their amazing journey from New Simon to Caruthersville and across the Boothspoint Bridge just as it fell kept her adrenalin level high.

Upon reaching the pair she said, "I talked to folks up the hill. They said things are a mess all over. People are streaming out of Dyersburg. The good news is that there may be a place we can eat in a couple of hours."

Alex stood. "Chris keeps blaming himself for this earthquake. He's got some crazy idea he caused it all with his prediction."

"Chris, you know that's not true. You just happened to be the one person who figured out when it would happen. Just because you were right doesn't make you guilty of anything."

Chris stared at the ground. "Well, whatever, I'm through with seismology." He stood and pulled his laptop computer from the backpack that had been saved from the wreckage of Alex's SUV. "Everything I learned is on this laptop: the model, the data, my notes. The world doesn't need it." He reached back and flung the computer out over the floodwaters. As the ripples from its splash washed the shore he said, "There. It's gone."

Alex and Tina stared in horror. Tina cried out, "Chris, why did you do that?"

Chris shook his bowed head and choked. "Someone told me it hurts to be a prophet. I don't remember who, but he was right. Now it doesn't hurt so much."

Tina stepped over and gave Chris a hug. She understood. She felt guilty to even be alive. "I wish I had something to throw, too." She paused then stepped back. "Come on. Let's go get something to eat. Then I think we should go back to Memphis. Maybe we can help the people there."

Seven hours had passed since the Tiptonville shock dropped the center of Caruthersville below the level of the river. The rising water had forced those few remaining inhabitants who could seek shelter into the upper reaches of the remains of the few buildings that still stood above water level.

The fracture zone continued to adjust the misbalance of forces as a variety of aftershocks pushed and pulled various portions of the seismic zone into lower energy configurations.

Another 6.6 magnitude aftershock originating 8.2 kilometers beneath the I-255 Bridge flattened the landscape even more.

"What are we going to do now, Jim?" Loretta asked the deputy with whom she shared the small second-floor room in the courthouse. Their haven had withstood the shaking, but the inhabitants of other rooms in the courthouse had retreated to makeshift rafts when the walls protecting them failed. Driven by the small but steady river current that flowed over the sunken levees, they moved off to the south, searching for some place to land.

"I figure we should just wait out this flood. I stashed a couple of gallon jugs of water and these cans of Spam up here early this morning. This concrete building is holding up real well. We'll just stay here and make ourselves comfortable. Pretty soon everyone else will be gone and we'll have the whole town to ourselves." His grin exposed the gap between his yellow front teeth and a crooked canine.

"Well, you haven't been very busy making up a raft like the others did, so I guess you'd better be right. We'll stay here. But do you really believe someone will come along and rescue us?" Loretta worried about how long she could remain in close quarters with her companion.

"Sure, there'll be someone soon. I expect the word has gotten out that Caruthersville has been flooded, and the State Police or National Guard will be along shortly. We could stay here for a couple of days, if necessary, but they'll be here before then."

"But what if they're in the same mess we're in? Don't you think that might keep them where they are?"

"Naw, this is the county seat. They've got to come by here and check out this place first. Don't worry. Everything will be all right. There ain't nothing' to worry about." He reached over and rubbed her knee.

Another small aftershock rocked their haven, but the building remained stable. Loretta reached over to take another cigarette from the deputy's pocket, lit it, and taking a long breath filled with smoke, closed her eyes and leaned against the concrete wall. She tried to ignore the touch and just daydream of rescue.

The *Bella Queen's* hull rested on the concrete and steel snag beneath the bridge. After almost an hour, the engine room chief climbed to the pilothouse and reported directly to Barney, "I've inspected every nook and cranny of this boat, inside and out. I found a slow leak around the rudder, but we shored that up. You're damned lucky you didn't cram the rudder up through the deck. But you ain't got no steering and the prop shafts are so bent they're unusable. It'll take six weeks in dry dock to fix this boat." Showing his disgust he leaned out the door and spat into the river.

The first mate asked, "What are we going to do now?"

Barney noticed the redness in Ralph's eyes from the shock of losing his son. "Well, I better tell Buddy Joe I can't find any way off this artificial reef. I know he told me not to head downriver, but my butt's in a sling and I need his help."

Barney called on the radio and reviewed in detail the condition of his boat. When he finished Buddy Joe asked, "Barney, it looks like there's room to get my towboat with its barges around the *Bella Queen* under the bridge. Do you agree?"

Barney estimated the distance between his boat and the south columns of the bridge. "There is room for the towboat and barges to pass on my port side provided you sail straight through. Yes, you can make it, but you better get it right the first time. So why do you want to get below me? You just going to leave us hanging here?"

"I thought of doing that, but no, I won't leave your passengers in that sort of predicament. It's just that I don't have any place to park these barges, so my plan is to go downriver, then motor back up to get in position beside you.

"I'll throw you a line as I go by, then I'll anchor west in the flood plains. When the surge lifts the *Queen* off the snag tonight, you should swing down on the towline. Then you can reel yourself in and tie up to the side of the barges as another part of my tow. Your anchor winch is strong enough to pull the *Queen* in against a seven-knot current, isn't it? We'll have to run the tow line to a cleat on the outer barge, so there's no good way for us to bring the line in."

Barney replied, "Yes, it should be able to do that. Else we'll get some able-bodied passengers down to help."

"Good. From there it's just a walk in the park to head on downriver to Memphis."

Barney thumbed the transmit button. "That sounds like a plan to me. By the latest estimate I've heard it'll be after midnight before we float free. So we have plenty of time to get ready."

"Get your crew ready now, Barney. There's a lot to do. As we go by, my first mate will throw a half-pound monkey-fist tied to a quarter-inch nylon heaving line from our starboard stern onto your bow. Your guys better catch it first time. Make sure you have your anchor winch clear. Then when you're ready to bring in the lead line, bring it in fast. It'll be tied to 600 yards of two-inch hawser. Your guys are going to have to pull it in and get it wrapped around your winch before we get too much of the line in the water. Think you can do that?"

Barney said, "Sure, but what are you going to do, jerk us off the snag?"

"The *Lady Bird* will be in full-reverse props as we go by, and I have to bring her to a complete stop before we pay out all the hawser. Then I'll move her and my barge tow back upriver on your starboard side and anchor near the bridge."

"So you want us to hang the tow line off the port side of our bow. That means when we lift loose, the *Queen* will swing around, right?"

"You got it. My boys will move the towline at this end down to an outer cleat on the portside barge next to the *Lady Bird*. That will be your docking point. Remember, you'll have to use your anchor winch to bring yourself in. I don't have any power down on the barges." Buddy Joe paused then continued. "I can't think of anything else. I move downriver in an hour and a half. My crew needs to retighten all our lines before then."

Barney interjected a comment. "By the way, Buddy Joe, there's an old black man in a boat tied to the bridge pier at the south columns. It looks like when the surge gets here he could be swept away."

"Yeah, Barney, that must be old Virgil. He's as ancient as the river and a friend of mine. He saved my life at least once and taught me how to fish. Do you think you can get him to row out to your ship?"

Barney keyed the mike to respond. "Sure. Anything for the guy who's going to save my butt; I'll get him on board."

Jud steered the johnboat into the channel to Mud Lake, now 10 feet deeper than before. "Lucky mom had the house built up the hill from the shore. This water's really backed up from that landslide at Fort Pillow."

"What're we going to do with all this stuff, Jud?" Sally Mae poked through the items in the bottom of the boat that they had collected from several fallen houses on the west bank of the river above Drive Cutoff. It looked like they had been to a garage sale.

"We'll hide 'em behind the house. We'll need to wait 'til things have settled down before we get in touch with Jake and get some money for 'em." The sun dropped behind the trees as he entered the lake proper. "Hope we kin find something to eat at the house. Why didn't you pick something out of the pantries at those houses? All you wanted to look for was jewelry and dresses. Ain't you got no sense?"

"I thought you said we'd get that big boat. I wanted to sleep in a nice bed tonight, and they probably have some good food on the boat. Why'd you pass it by?"

"I seen the old man talking on the radio, so if we had set on them right then they could have called for help, or identified us. I figure they'll hang around. Besides, this johnboat's not fast enough to catch 'em if they take off."

"We have the beer you left at the house. Maybe I kin find some crackers and cheese."

Jud steered the boat with the outboard motor to avoid a fallen tree. "This boat's too small and slow with us in it. I think there's a couple of jet skis up at the Purdy's place. They said something about not coming up this weekend, so the skis are there for the taking. They'll be fast enough to catch the big boat and we can pull the johnboat along behind."

Sally Mae nodded her head in agreement. "I'd like a beer. It makes me tingly."

"That's what beer does to someone who's only 12, didn't you know that. You need to be sure you don't drink so much you pass out like last time." Jud grinned at his sister. "I guess it's my job to keep you awake, at least for a while."

Buddy Joe commanded the engine room to provide quarter-speed-ahead as he maneuvered the barges to come around for the approach to the bridge, half a mile downstream.

The radio chirped. "Buddy Joe, most of the room's on my port side, so come down near the south shore. My maps show the channel remains deepest up next to the bank along there anyway."

Buddy Joe thumbed the microphone. "Sure thing, Barney. I already figured that out." He turned the rudder hard to starboard and commanded full power, pushing the back end of the tow parallel to the south bank. "Reverse props. Let's get this tow aimed straight at our target before we go sailing into that hole." He swung the rudder back to port and brought the tow to rest relative to the land.

The radio chirped again. "Looking right good there, Buddy Joe. Maybe I could get you a job as a pilot on this river, the way you're driving that old towboat."

"Props reverse, quarter-speed." Buddy Joe thumbed the microphone again. "Quit pestering me, Barney, or I'll ram your ship." His chuckle masked the tension he felt about the delicate maneuver under the bridge. "I hope you realize I don't normally have to pass ships anchored in the middle of the channel under a bridge. I really should report you to the Army Corps for violation of the river rules."

The sound of the towboat grew louder and louder as it neared the bridge. Virgil squatted in his boat next to the pier and watched the barges approach the bridge from the northeast. "Well, that tourist boat must be going to get some help from the *Lady Bird.* I see the *Lady Bird's* sticking to this side of the channel." He grinned with pride. "Buddy Joe will miss the snag by lots. He's learned a lot since he was a kid."

Virgil watched the white towboat maneuver the back end of the tow to thread the barges through the space between the riverboat and the big columns supporting the bridge. The first of the barges glided by 20 feet away from his rowboat, the river current moving several knots faster downriver, making the smallest of ripples in the water. The barges blocked his view of the riverboat. He counted them as they went by. "… Three … Four."

The towboat sailed past the bridge column, already working to steer the barges ahead of it starboard. Virgil could see the tall burly redhead dressed in a plaid shirt, dark blue overalls and a blue cap pick up the battery-powered megaphone and shout to shore from the wheel house atop his towboat.

"Virgil, Virgil, listen to me. The locks broke on Kentucky Dam, and there's a flood on its way down the river. It's raising the level six or seven feet and should be here about midnight. You got six hours to get out of there or you'll drown. Didn't the *Bella Queen* tell you to get away from that pier? Do you understand?"

Virgil stood and waved to the towboat from the rocking rowboat. He shouted back, "I heard you. Just wanted to be sure you steered your towboat right and made it through okay."

Buddy Joe shook his head in dismay. He thought Virgil would have moved, but now there were other things to worry about.

When the barges glided past the riverboat and the towboat came nearly even with the stern of the stricken ship, Buddy Joe radioed to ask, "Barney, are your crewmen ready at your bow to receive the tow rope?" He looked across the narrow space between the two boats. "We only get one chance."

"Ready when you throw it to us."

Buddy Joe shouted through the bullhorn to Jeff at the stern of the towboat. "Throw the heaving line. Stand aside and watch out the lines don't snarl." Jeff swung the half-pound lead ball two times in a circle over his head then heaved it with its trailing rope up and over toward the *Bella Queen*.

The riverboat crew caught the ball on the first bounce and wrapped the small rope twice around the forward anchor winch. As the line became taut, the two-inch hawser began to unwind from the coil on the deck of the towboat and slide into the water. The crew on the *Bella Queen* had already started winching in the heaving line to bring the end of the hawser to their deck. Dragging the dripping heavy rope to the deck, they dropped it through an anchor chock then cinched it to a large cleat for safety before winding a couple of loops around the anchor winch.

"Your stern's well clear of the *Queen*," Barney called on the radio.

Buddy Joe moved the rudder to propel the towboat hard starboard. He called to the engine room. "Going to full-speed reverse with hard left rudder." The *Lady Bird Jamison* shuddered as it sucked the water under its keel and fought the current to stop its forward progress. He prayed again that the steel ropes holding the barges together and to the towboat would not snap. This maneuver again put them under heavy strain, though not as much as the earthquake.

The towboat labored to bring the barges to rest relative to the bank and then moved them north out of the main current. The towrope billowed out to the side in an ever-growing arc between the towboat and riverboat.

It took 30 minutes for the towboat and barges to stop and move 100 yards crosscurrent toward the north bank. The sun had set and growing darkness made maneuvering even more difficult. It took another 20 minutes to move back to within 100 feet of the bridge.

"Drop the stern anchors," Buddy Joe commanded his crew. The twin anchors dropped to the river bottom 18 feet below and Buddy Joe eased the towboat forward until the anchor flukes planted themselves into the mud and held the towboat in place.

"Barney, does your crew have the tow rope tied in yet?"

"We're secure."

Buddy Joe called to his crew to secure the hawser to the outer cleat on the barge then called back. "Take up the slack."

Barney called to his crew on deck, "Bring in the slack in the tow rope." He watched as the rope winch shortened the towline between the towboat and the *Bella Queen* to the point of optimal slack and the crew cinched the rope to a cleat.

"Okay, Buddy Joe, we're secure at this end."

Buddy Joe breathed a sign of relief. "Fine. I guess now we can wait for the surge to come and kick your butt loose from that snag." Buddy Joe looked at his watch. They had taken two and a half hours to complete the maneuver.

"Sure thing, Buddy Joe. I ain't got nothing else to do for the time being—except to say a mighty big thank you. And that's from all my passengers and crew as well. God bless you."

— 11 —
INTERLUDE

The speakers sounded throughout the *Bella Queen*, sending the Captain's message to everyone on board.

"Ladies and gentlemen, this is Captain Ruggs with a status report. We are still snagged on the bridge debris, but now we have a line from the *Lady Bird Jamison* towboat connected to our boat, so we are safe from being washed away. We expect the flood surge from the Ohio River after midnight, and when it reaches here we will be lifted free from the snag.

"Captain Buddy Joe Simpson of the *Lady Bird Jamison* will lash the *Bella Queen* to his barges and give us a tow down the river in the morning. It will take a day or so to make it downriver, and we plan to dock at Memphis or some other suitable location, depending on what we find when we reach that area.

"Those of you with cell phones know that no phone service is available. Some of you have asked to send or receive personal messages, but we have no communications other than the emergency marine radio channels, and those are fully allocated. We cannot send or receive personal messages. Sorry. We will resume that service as soon as possible.

"Refugees from Caruthersville have been assigned to every stateroom as well as the game rooms. We have no washing facilities, so please take care of your own bed linens. Those of you who are hardy enough may enjoy sleeping on the deck tonight, though the forecast is for thunderstorms in the early evening.

"In the meantime, we are on limited water rations. Do not use tap water for any purpose other than drinking. Toilets are being flushed with river water from the buckets in the heads. Please refill the buckets when you use them. We have fuel enough to run the generators for required uses, but do not use the lights excessively. We have enough food for all passenger, crew, and refugees for three days, maybe longer after receiving a giant catfish from our friend Virgil. In celebration of Virgil's good fortune the chef will be serving chunks of deep-fried catfish and Cajun red beans and rice at seven o'clock tonight.

"For those of you who want to keep up with the news, the satellite TV in the lounge area will be turned on every two hours for 10 minutes so you can watch your favorite news show, just so long as it's NWS news. Sorry, but there are no sports shows tonight.

"Please bear with us, folks. There has been a terrible tragedy around us, and we on this boat are relatively well off for the moment. With God's help, we'll be able to land you safely at a port of call in a couple of days. If you have questions and cannot find me, please direct them to the purser. Thank you."

"My daughter is staying with friends east of St. Louis. I hope they're all okay." Lynn leaned against the railing with Alta Warren outside the cabin they now shared and told of her life and family. "It sounds like St. Louis is heavily damaged, but I don't know about Scott Air Force Base. It's mostly flat around there and there aren't that many tall buildings, so probably everything should be okay."

"How old is your daughter?"

"She's 12. Her name is Shirlea. She's just beginning to be interested in boys, and I'm a little concerned about who she's going to become in the next year or so."

Alta laughed, "You should be concerned. I went wild when I turned that age and my parents were ready to throw me out of the house. It's a hard time of life to be in. How do you think she'll take this earthquake?"

"Well, she has a pretty level head, so she should do all right. I'm not sure what my being away will do to her, but I have been away before for several weeks at a time on temporary duty for the Air Force. So she's accustomed to being on her own."

"What about you and Ron? Are you going to marry him?" Alta asked.

Lynn blushed and laughed. "I don't know. He's a really nice guy, but he hasn't asked me yet. I think he's still smarting from his previous marriage. I know Shirlea really likes him."

"Well, when you see a good man, don't lose him."

"Oh, I think I can get him on the line." Her face glowed, and then she could not hold back her feelings. "Keep it a secret, but I've decided that's what I want. After what we've been through he's become somewhat protective, so that's a good sign. I want him to be with me. I'd like for him to be the father of my kids."

Alta squeezed Lynn's hand. "It seemed obvious that you've fallen for the guy. He seems like a good man and he cares about you. He's also pretty good in this kind of situation. Hang on to him. I know he'll be good for you and Shirlea."

Lynn turned and looked into the older woman's eyes. "You know, it's so strange that even though the world around us has been destroyed, we can stand here and plan for the future."

Alta bowed her head. "You have to. Tim and I have lost a son and daughter-in-law, but we gained a family. It's at times like these that we have to look ahead."

Passengers and refugees filled the *Bella Queen* lounge to watch the large screen TV. The NWS anchor Harold Owens looked into the camera and continued with his report, "It is now evening along the New Madrid earthquake zone, but the seismologists tell us all is not quiet. Aftershocks continue to pound the area.

"For the most part any area within 200 miles of the fault zone is without power. Pilots flying over the earthquake zone report that they see very few lights, and most of those must be fires set intentionally or that have not yet been controlled.

"President LaPorte has scheduled a cabinet meeting for tomorrow morning. He will discuss his declaration of Martial Law that is to be implemented under the command of the Army and his plans for providing aid to the regions that have been declared disaster areas.

"Aid is pouring in from Europe, South America, Mexico, and Canada. The Red Cross is organizing supplies to be shipped into the earthquake areas as soon as that can be arranged. However, transportation into the stricken area is proving to be a major problem. Most bridges over rivers and small streams in the earthquake zone have fallen. The rail system has been destroyed. Overpasses on the major highways are down and they are blocking the freeways and secondary roads. Flooding in

Arkansas, Mississippi, and Louisiana has covered many roads, restricting any movement from the south. And authorities fear the river barge transport system south of St. Louis has been ruined.

"We now have a special report from Ted Krantz regarding the transportation system in the five-state area surrounding the New Madrid fault zone. Ted?"

Ted Krantz appeared on the screen with a backdrop of an interstate overpass. "Harold, I am standing next to I-40 exit 159, east of Little Rock, Arkansas. There is some traffic flowing on this interstate, but that is not true farther east of Little Rock, north of Vicksburg, west of Knoxville and Cincinnati, and south of St. Louis. Much of the interstate highway system in that area is no longer passable. Overpasses and bridges have tumbled from the shaking they suffered from Saturday morning's earthquake and from the aftershocks that continue to rattle the region.

"We have a report from the U.S. Department of Transportation issued late this afternoon. They estimate that as many as 1,000 major bridges collapsed in seven states because of the earthquakes. Another 1,500 are severely damaged.

The TV screen switched to an aerial view of a four-lane highway slicing through a canopy of green trees spotted with large fields. "This video highlights some of the damage we saw earlier this afternoon on a flight from Oklahoma City to Little Rock and the Memphis area, north to St. Louis and back."

"The roads around Little Rock all appear to be open, though there is limited traffic heading east. The Army has established roadblocks at the edge of the region they expect will be flooded during the next few days.

"As we flew over Marked Tree we could see several areas of flooding around the St. Francis River and no indication that any bridges remained across the river itself. We were told this area is near the southern end of the earthquake fracture.

"Coming into West Memphis, you see that several overpasses are down around the interchange between I-40 and I-55. As you can see there are groups of cars, trucks, and people gathered atop most of these overpasses. These are the highest points of land in the surrounding area and offer refuge from the floods. We counted at least nine bridges down along this section of the Interstate.

"Apart from the collapsed Pyramid, one of the most striking failures is across the Mississippi in Memphis at the east end of the I-40 Bridge. It appears a huge earth slide has moved the bridge and Mud Island away from the shoreline, dropping the bridge connectors into the Wolf River. In the city of Memphis, most elevated portions of the interstate system are down. We saw no traffic moving on any portion of these freeways.

"We turned south over Memphis but could see little beneath the cloud of smoke still billowing up from President's Island. We assume the bridge problem is the same all over. There is some obvious damage to the I-55 Bridge across the Mississippi and the two accompanying railroad bridges, but it does not look as serious as damage to the north.

"We turned north to follow I-55 into Arkansas. Surprisingly, it appeared that almost every overpass had collapsed along this route. Maybe it was just a matter of being close to the fault, but one wonders about their construction. A major retrofit program last year should have prevented this level of damage.

"There are numerous levee breaks along the Arkansas shore of the Mississippi River, and water is spreading across the land. A giant slide at Fort Pillow appears to have possibly blocked the main channel of the river.

"The damage to the I-255 Bridge across the Mississippi south of Caruthersville is total. In addition to the collapse of a portion of the main roadway over the channel, all causeways connecting that bridge to the opposite shores are now covered by the floods."

Several of the people in the *Bella Queen* lounge recognized the area. "Hey, there we are. See, the towboat was above the bridge when they filmed this."

The camera passed over the remains of Caruthersville. "There is little left standing in Caruthersville. The area is totally flooded. In fact, flooding is evident along the entire course of the river, including the area around New Madrid.

"We continued up to Cairo, Illinois. The two bridges at the tip of land where the Ohio and Mississippi Rivers join suffered severe damage on their short causeways, but they are still standing. This area is under flood water from the break at the Kentucky Dam, and the town of Cairo has all but disappeared."

The two bridges stood like lone sentinels to mark the channels of the two mighty rivers. No evidence of Fort Defiance could be seen. The remains of a few buildings peeked above the waters at Cairo, three miles above the bridges.

"As we made our way up the Mississippi River, we saw fewer and fewer fallen overpasses. The new bridge at Cape Girardeau seems to be useable; we could see some light traffic on it. When we reached the St. Louis area and circled the city, we saw that several overpasses in their freeway system had fallen, but the damage did not appear to be nearly as severe as what we saw in Memphis."

Harold asked, "Ted, did you see any river traffic?"

"There were some barges and towboats along the river, but for the most part they seemed to be anchored, probably awaiting word about the condition of the river. I understand all traffic has been stopped on the river by order of the Army Corps."

The camera focused on different bridges, zooming in when it found some area of damage. Ted returned to his discourse. "That is just some of the damage caused by this earthquake. The worst of it is found along the trace of the fracture, but some roads hundreds of miles away have been affected. In many cases, it will take months or even years before the transportation system can be restored to its preeminence.

"This is Ted Krantz, reporting from Little Rock."

The catfish dinner proved a hit with passengers and refugees. The cook sliced the big fish into chunks, dredged the chunks in a mixture of cornmeal and salt and pepper and fried them in deep fat. No one worried about the bones as every bone in the big fish was at least a quarter of an inch in diameter.

Ron and Lynn sat around the dining table with Virgil and the Warrens, telling each other of the various experiences that had brought them together.

"Captain Lynn Browne and I have been working on contingency plans for Federal Express and the U.S. Air Force for disasters like this. Even though our directive was to focus mainly on terrorists, we did consider the effects of an earthquake."

Lynn picked up the conversation. "This was a contingency plan between Federal Express and the U.S. Air Force Air Material Command.

Ron retired as an Air Force Major and C-17 pilot. The report's primary concern focused on what to do if the Memphis or Indianapolis airports went down and whether Federal Express could move some of its operations to Scott Air Force Base east of St. Louis."

Tim Warren asked, "What did you conclude? Did you consider this big an earthquake?"

Ron shook his head. "Not really. We heard on the grapevine that someone at the University of Memphis had made a prediction for a big one this weekend, but my boss at Federal Express insisted that we not include anything about earthquake predictions in the report.

"We did some estimates for damage from a 6.5 earthquake and concluded that was not enough to stop operations in Memphis. The TV said this one measured magnitude 7.9, so it probably tore the airport up pretty badly. This is far worse than anything terrorists could have done. Probably put corporate headquarters out of business."

Lynn added, "Our conclusions were that there had not been sufficient planning for how to handle things, especially with regard to recovery, in the event of a moderate earthquake, let alone a really big earthquake."

Tim asked, "What do you think is going to happen now? Will the Air Force and Army be able to help this area?"

Ron expressed his confidence. "I'm sure the President and Governors are busy mobilizing the National Guard and Armed Forces already. The Air Force could start an airlift to bring in supplies and the Guard can move in to help maintain order. Help should be pouring into this area already."

Virgil shook his head in disagreement. "I don't think the gov'ment is going to do much."

Ron looked surprised and asked, "Why do you say that?"

Virgil squinted. "My pappy and mama was just married when they had the big flood of '27 in Lou'siana. Pappy said the gov'ment never did anything for his folks.

"He sat there on top o' their shack, he an' Ma – sat there fo' four days. But nobody came near 'em. Finally, they made a raft out of parts of the house and paddled for two more days 'fore they found a place to go where there was people who'd give 'em food. They had to drink river water. Ma got sick along the way and almost died."

"Pappy was forced to work on the levee gangs. He didn't have no choice if he was gonna get any food. Finally, when the gov'ment wasn't looking, he and ma sneaked onto a boat and left. Didn't seem to be no reason to stay down there in Lou'siana. They didn't have no family, the farm they was share-cropping was gone, and there shore wasn't no gov'ment to take care o' anybody."

Ron looked at the old man. "How old are you, Virgil?"

"I reckon I was a gleam in my mama's eye 'bout that time. I was born in Memphis the end of '27."

"But Virgil, don't you think things have changed since then? Don't you think the state and federal governments are better prepared now?"

"Ya'll just wait an' see what I say. The governors in Nashville and Little Rock and Jackson are all busy taking care of their own. The federal gov'ment just won't have time to worry about us folks along the river. They'll get to arguing on what to do and how to do it and never get around to helping. We're just too far away, like in '27."

"Ahoy there." Clarence Forsyth steered his 15-foot outboard craft in the darkness toward the anchored towboat until it bumped into the bigger hull. He called. "Did you folks lose a deckhand named Ricky and a passenger named Joe?"

The crewman on watch came to the starboard side of the towboat and shown a light on the small, crowded boat. He asked, "Where'd you come from?"

Clarence called back. "My wife and I and a neighbor boy came upriver from the Obion River where we live. Things are really messed up down there, and when I heard on the river CB about the dam breaking and the river rising I piled my family into the boat. The break in the levees totally flooded our place—not that it mattered much. It was all wreckage anyway. I was hoping we might find something better up this way.

"What's this about a deckhand?"

"We picked up these two pilgrims hanging in a tree downriver a couple of miles. We were about to tie up to a tree branch when we found them, and they said there was a big boat up this way. We had run out of water, and I'm about out of gas, and this boat is really crowded with those two. They're pretty sick from swallowing river water. So I decided to head upriver to find you guys."

"Well, the two refugees you rescued must be from the *Bella Queen*. She's snagged under the bridge until we haul her in later tonight."

"Can we tie alongside and stay with your boat? We have a neighbor kid and he needs something to eat. We could also use some drinking water."

"Sure, move your boat down and hitch a line to a stanchion on this barge. Climb up and bring everybody to the towboat. We've got some leftovers. Your boat may get bounced around a bit, but that's the safest place to dock."

"Thanks." Clarence nestled his small boat alongside the barge as directed. With the help of other refugees and crewmembers, he transferred the two bedraggled survivors from the *Bella Queen*, his wife and the child from his craft up the four-foot high sides of the barge. Then with a line tied to a stanchion on the large craft, he cut his engine and let his boat drift in the current next to the bigger craft.

A driving rain was falling from a thunderstorm crossing from Arkansas into Tennessee. To try to stay dry, Ron and Lynn sat as far back into the deck chairs as they could, against the wall and under the overhang of the deck above them.

A brilliant streak of lightening reached down to the grove of trees on the bank and pulsed and pulsed. Feathery streamers of electricity spread out to the side, searching for other paths to the ground.

The crack of thunder came at the same time, so loud they could feel it hit their bodies.

"Wow, that was a close one, Ron." Lynn laughed with excitement. "I always get a thrill from watching and hearing the pyrotechnics of thunderstorms. Do you think this riverboat is safe?"

Ron felt the hair lay back down on his forearm. "Sure hope so. They probably have lightning rods on top with good ground cables into the water. We're sitting ducks out here in a storm."

Ron shoved farther back in his chair. "Boy, this rain will make it miserable in the towns that have been hit by the earthquake. At least it should put out some of the fires." He watched as sheets of falling water reduced the visibility of the shoreline to zero. At times, the wind coming from the other side of the boat whipped the rain into horizontal sheets.

Lynn snuggled her head into Ron's shoulder, hugging his arm and looking to the side out across the water. "I love you."

Ron smiled. There it was again, that spontaneous expression of love. He found it addictive, maybe even required from time to time.

Reaching across with his free hand he rubbed Lynn's shoulder. "I love you, too, Lynn. I love you a lot. More than I ever planned to."

She sat up and looked him in the eye. "Oh. You mean you had some plans. Pray what, kind sir, were your plans?" She grinned and pecked him on the cheek.

"Well, my plans were to have a good fling and lots of fun. Now I guess I've changed my tune."

"Oh, I'm sorry. I wanted to have lots of fun, too."

"I still want to have the fun, but I guess I want more. I guess I want you around me all the time." He paused and she waited, saying nothing. "Lynn, if I asked you to marry me would you consider it?"

She waited several more seconds. He began to worry, to wonder. "Well, of course I would have to consider it. Such a matter requires a great deal of consideration. I'd have to think for quite some time about that issue, what with all the extenuating circumstances and everything else going on, you know. It would require an awful lot of thought, and it could take a terribly long time to consider it. I suppose, maybe, but I don't know, hmmm."

Ron squirmed in his chair. "Oh, shut up, you sound like an Air Force Officer. Will you marry me?"

"I thought you'd never ask. Yes." The wind turned and blew the rain under the overhang. The pair clung to each other in rapt ecstasy, oblivious to the drenching downpour the deck chairs were receiving from the evening thunderstorm.

— 12 —

The Surge

Buddy Joe tried to nap as he leaned his stool back against the bulkhead of the wheel house. His head bobbed and fell forward. His eyes closed and his mind slipped into a dream.

His gloved hands gripped the steering wheel of the black and gold racecar as it rocketed around the track. Concrete walls whipped by in a blur. The roar of the engine pounded his ears as he pressed on the accelerator. The steering wheel vibrated harder as he forced it to the left, fighting to stay in the groove. He felt the rear tires let go. Like in a slow-motion dream the car drifted toward the wall. He slammed on the brake and turned hard right, trying to push the car back to the track.

"You're in pit row. You're going too fast." A voice yelled into his earphone. The car struck the wall and bounced back, shaking and sliding toward the man on the jack who was changing the outer tires of the car in the next pit.

Motion slowed to the extreme as Buddy Joe saw his car lift and the man on the jack look up. He recognized the face and screamed. "Jimmy, get out of the way, Jimmy."

The bright flash of an explosion blinded his eyes. "No. No." A crackling roar burst on his eardrums. His mind snapped, searching for reality.

A tremendous clap of thunder obscured the dull roar of the aftershock rocking the towboat. Buddy Joe awoke with a start, confused. Where was he, what was happening?

The towboat tossed with the chop. Buddy Joe pushed off the wall, now fully awake, and grabbed the edge of the wheel house door. He looked out and around. Something had changed. Something felt different, something the lightning and shaker had tried to warn him of.

"Jeff," he hollered to the deck below. "Quick. Shine your light on the *Bella Queen*."

Jeff switched on the powerful floodlight and illuminated the snagged boat.

Buddy Joe watched a moment, and then yelled again. "The *Bella Queen's* moving. It's moving upstream."

"Cap'n, I think it's our anchors dragging. That was a pretty big aftershock and the mud must be pretty sloppy down there in the bottom of the river."

As the roar ended and the chop subsided from the surface of the river, Buddy Joe studied the situation. "You must be right." The cramp in his back subsided and he eased his grip on the railing outside the wheel house.

"Jeff, it looks like that temblor moved us 50 feet downstream, but there's still enough slack in the towline to handle that. No need to re-anchor the towboat unless we slip a lot more. Keep an eye on things."

The thunderstorm moved in with a vengeance, lightning and thunder blasting the banks and bridge. Buddy Joe stepped back into the wheel house and returned to his stool. The cooling rain helped, but the tension and not knowing what would happen again kept him awake.

His mind drifted to Ellen, his wife, and he worried how she might be doing. Barney had told him over the radio of some of the satellite TV news, about how little was known of the damage downriver around Memphis, but it had been enough for him to know she was not out of harm's way.

He speculated about the rest of the *Lady Bird* fleet. Three boats and fleet headquarters on the Loosahatchie, all to the south, had never checked in on the radio network. He wondered what happened to Paul Taylor on the *Lady Jane Wilson* down in Osceola. Paul had not responded to repeated radio calls since the first big aftershock and the tidal wave that morning.

Tired and drowsy, he waited in the wheel house of the towboat, unhappy he could not sleep but afraid that he might. Rain pounded on the window and roof of the small room. His head bobbed and fell forward. At last his eyes closed and his mind slipped into another dream.

Charlie leaned back in the canvas-covered Captain's chair, his feet propped on a tackle box. Downstairs in the stateroom he could hear Sylvie and the girls giggling, playing some kind of game.

The *Amanda Blair* lay two miles upstream from the giant earthen dam, out of the sound and grasp of the rapids just above Fulton. When he anchored, his GPS had told him that his boat was at the head of Plum Point Reach, next to the eastern shoreline. The water level had risen to the point where only a few treetops showed in the fading sunset, and they were hardly enough to tell where the flood plain should be.

Charlie considered the quiet to be deafening, matching the totally black sky. Clouds covered the stars and there were no reflections anywhere along the horizon, except for an occasional distant flash of lightning.

He had not thought about it before, but with the absence of lights along the horizon, one more mark of civilization had disappeared. Only the LED lights of the instruments were there to provide enough light to make out the boat controls.

He played with the dial of the satellite radio boom box he had purchased the previous week. One of the newer toys of technology, the radio received the satellite's broadcast signal anywhere.

Every local radio station tower must have been wiped out in the shaking, or else they had lost all power and could not transmit. This advance in technology at least allowed him to listen to the news.

The news was not good. Seismologists pegged the earthquake at magnitude 7.9 and there had been another 7.4 temblor shortly afterward. That must have been the one that brought down the bluffs. He heard lots of talk about organizing help and sending convoys into the stricken area, but he heard no news of anyone getting there.

Thunder rolled in from the north. Areas up around Caruthersville must be having a strong storm. The way the lightning seemed to be moving, he would probably be drenched in the near future.

The girls said something, and then he heard Sylvie laughing, almost hysterically. Charlie felt good that she could be happy again. She had at last found someone to replace her granddaughters. He smiled in the darkness for the first time in a long time.

The radio call from Captain Dennis Bugler of the *Lady Janet Quayle* near Cape Girardeau had aroused Buddy Joe from his fitful nap. After a short discussion he pressed the button on the radio to resume scanning.

Jeff had stepped into the wheel house in the midst of the conversation. Afterwards he asked, "Cap'n Bugler said the Army Corps have stopped all traffic on both rivers until further notice and don't have any ideas about when it'll reopen. Does that mean we can't move from here?"

Buddy Joe sipped at the cup of coffee Jeff had brought to him. "Thanks for remembering the ice cube this time." He took another sip. "As far as I'm concerned, we're in a state of emergency and will do whatever we find necessary. I think they're concerned about traffic outside the region that has been damaged."

"What about the President declaring Martial Law?"

"Well, it doesn't surprise me much, but it surprises me that the Army will be taking over and not the National Guard. It sure doesn't sound like anybody's getting their ass in gear to fix things back up. Denny said he had heard no word about when there'll be any relief crews on the river."

"Cap'n, what should they be doing? Do they even know?"

Buddy Joe took another sip of his coffee and moved over to scratch his back on the door jam of the wheel house. "I doubt it. The time for rescue is nearly over. Ninety-eight percent of those rescued are saved in the first 12 hours. No one has really thought about relief and recovery.

"I talked with a friend in Emergency Management in Memphis, and he said they practice rescue procedures, but no one wants to take the time to figure out how they would recover from something like this. So no, I don't think anyone really has a clue of what to do now.

"From what Denny says the *Lady Bird* must have been sitting directly on top of the fault, and there's major damage north and south of us. It may be that communications are so bad no one really knows how bad things really are. We know places like Caruthersville are toast because we're here."

Buddy Joe leaned out the door of the wheel house to look into the dark sky. Since the clouds had passed, the stars formed a canopy over the river, more than he remembered ever seeing. There were no lights along the horizon to compete, save those on the two boats. A few small glimmers flashed now and then off to the south, the remnants of a dying thunderstorm.

"You know, Jimmy, I have a bad feeling about all this. I don't think we've seen the worst of things yet."

Jeff shifted his feet. "That's twice you've called me Jimmy, Cap'n. Who's Jimmy?"

Buddy Joe blushed, thankful for the darkness. "Oh, I'm sorry, Jeff. My mind is wandering." He paused and coughed. "Jimmy was my son. You remind me a lot … we lost him in a racing accident. He's gone now." He stared into the pitch-blackness around the boat and let his tears flow without bothering to blink. With this earthquake the whole world now knew and understood how he felt. "So many are gone now. So many."

From time to time voice traffic on the marine radio told of disasters up and down the mighty river, of broken levees and burning towns and chemical spills into the waters. Calls for help came often but Buddy Joe heard few offers to do anything about them.

From time to time he felt the towboat sway from side to side at the end of its anchor chains, a sure sign of another aftershock. His eyes drooped. Once again his head bobbed and began to tilt forward. A sudden burst of chatter from the radio interrupted his reverie.

"… river broke through the levee south of Bessie. I repeat the river has broken through at Bessie." Buddy Joe reached over to turn up the volume.

"This is Steve Pauli. I'm stranded here at Bessie with my daughter, Juliana. We were farther south when the big earthquake hit, and it almost cut through the levee then, but now the river's broken through with a roar. When that flood surge reached here it poured over the top and the water's cutting a whole new channel. It must be over a half-mile wide.

There was a pause, and then the voice continued, "I just saw a couple of cars on the south side of the break and they look like they're falling into the water. Yes, there goes another one. Oh God, there's a whole line of cars over there that's dropping into the water."

Buddy Joe heard another pause then, "Oh no. My God. This levee we're on is starting to move. The river's pushing it to the side. Juliana, watch out. No." The voice rose to a scream. "Oh my God …"

Buddy Joe heard a burst of static and then the channel went quiet. The radio resumed scanning. He stood and reached out to return to the frequency of the previous transmission. Still quiet. He repeated the expletive he had heard over the air in empathy. "Oh my God. That means a waterfall has just replaced nineteen miles of river around the New Madrid bend."

He waited a minute to hear more from whoever had been speaking. "Steve Pauli," the voice had said, but the channel stayed quiet.

Buddy Joe called and searched through the radio bands seeking more information about the break. He found a couple of other listeners who had heard the same message, but he could find no one at the site of the supposed break. Someone said he would try to contact the Army Corps of Engineers and authorities at Tiptonville.

Flipping to the frequency used by his sister ships he called for them to wake up. He had instructed Barney to monitor the same channel from the pilothouse of the *Bella Queen*.

"I just heard a partial report on the radio that the river has cut through Bessie's neck. I was unable to get confirmation. If that has happened, then we can expect the surge maybe 90 minutes earlier than predicted. And there will probably be a tremendous amount of debris in the river. We'll have to wait and see."

Barney responded to the news. "What do you think, Buddy Joe? Is this going to create a problem for us?"

"Well, Barney. If the surge comes through quicker it might help in lifting your boat loose from the snag. We'll just have to keep close watch that we don't get caught with debris tangled in the towline. And once you swing downriver be prepared to drop anchor in case the towline fouls and we have to cut it."

Buddy Joe looked at his watch again; it was only a little after 10:00. He spoke to the darkness. "Steve Pauli, now you are gone, too."

There was no evening glow, and it made the river seem even quieter. Freddy leaned back against the stanchion on the barge deck and cuddled Paula in his arms. She moved their daughter to her other breast. The baby drained milk from her mother's breast as fast as it could.

Freddy fidgeted and once again stood to look around. "I ain't seen nobody along the river for a long time now. Usually there's someone around the houses along the shore, but there's not a light to be seen." He kept hoping to see some signs of civilization on their ride down the flooded Mississippi.

Paula said, "Maybe they all headed into town."

Freddy figured Paula must be searching for some non-threatening answer for the lack of life. "Maybe you're right, but it bothers me. With nobody around, it could mean we're moving into places with more damage than upriver, though that's hard to believe." He sat and leaned his head back against the post and closed his eyes for a moment.

Holding his breath, Freddy noticed a low-pitched rumble similar to the noise before an aftershock, but it seemed more constant, steadier. It seemed to be growing louder.

"Paula, do you hear a low roar?"

Paula took her attention away from her daughter and listened. "I guess so. It's awfully quiet, but I do hear something. Do you have any idea where it is?"

"It's hard to tell, but with it getting louder I think it must be coming from downriver."

Over the next five minutes the sound grew louder and louder. In time it became a distinct growl.

"Paula, I know it's crazy, but it sounds like a waterfall. I heard something like that when my folks took us to Colorado one time."

The pair could feel the barge shift and start to swing, like it was moving into a quicker current. Their apprehension increased as the sound grew from a rumble into a roar.

Freddy stood and looked ahead. "We're moving into a faster current. That's what's making the barge rock from side to side and jerk around." Freddy saw the headlight of a car on the shore ahead of them. Using that as a reference he could see that the barge moved very fast, faster than he had ever seen the river run.

"Hey! You on shore! Is anyone there?" Freddy stood and yelled at the headlights as they whisked by.

An answering call came from the shore. He saw no one, but he heard a voice above the sound of the river. "Hello. Who's there?"

"We're on a barge drifting by. Where are we? What's the noise?"

"You're just above Bessie, but the river's cut through. Watch out for the waterfalls just ..." The voice disappeared, lost in the growing roar.

The barge twisted and jerked harder as it washed back and forth in the funnel of the water chute.

Freddy dropped to his knees next to Paula as the roar became deafening. He wrapped his arms around her and around the stanchion and shouted, "Hang on, we're going into some big rapids. The man said something about a waterfall."

"Oh, Freddy, I'm scared!" Paula clung with both arms to her daughter and buried her face into the arms of her lover and protector.

Jim became more and more insistent as Loretta lay on her back on the hard wooden floor. The darkness covered their activities, not that it mattered, for no one else remained in town to pay attention to Jim's wild lovemaking.

Loretta remained placid, accepting Jim's exertions without emotion. She had done this so many times before for various kinds of payment. Jim was just another trick so she would have a place to spend the night.

A cold sensation licked at her heel. "Jim, what's that?" Loretta pushed against the straining body above her as she felt another cold wash of water across her foot. "Jim, stop. I feel water on my foot. There's water on the floor."

Jim would not to be distracted. "Oh, baby, that's nice. Oh, baby, that feels so good." He pushed up on his hands.

"Jim, something's wrong. Get off me, you damned fool." Loretta shoved against the naked man and this time rolled to the side, throwing Jim the other way.

"What the hell are you doing? I wasn't through, bitch. What's the matter with you?"

"There's water on the damned floor. I felt it on my foot. This building must be sinking. Where's the flashlight. Shine it down there." Loretta pulled her clothes to her body and looked out the broken window into the blackness. Clouds again covered the stars and no light came from anywhere else in the city. She could see nothing.

"What are you talking about?" He found the flashlight and shined the meager beam into the abyss beyond their building. He wagged it around several times over the blackness of the floor before he realized he saw water inching across the lower edge of the linoleum in their small room.

"Hey, you're right. There's water right up to this floor. Where'd that come from?" Jim stood so he could shine the light down into the water. He and Loretta watched in horror as the water continued to creep up the sloping linoleum. The building had sunk further, or else the water had risen.

Buddy Joe stared in the darkness at the luminous dial of his watch and spoke to no one in particular. He just vented his frustrations.

"Twenty minutes to twelve. Damn, still an hour to go." He leaned forward and peered out the window of the darkened wheel house wondering when the next aftershock would rock the boat.

Jeff's voice called from the deck below. "Cap'n, I think I see a big sawyer coming this way, headed for the towline." As the lookout, he had been scanning the upstream waters of the river with a power light for the past 30 minutes, watching for just that situation.

Buddy Joe relayed the warning by radio to the *Bella Queen*. "Barney, wake up. Jeff reports a sawyer headed for the towline. Get your men on deck. We're going to have to play this one by ear."

Jeff focused the beam of light on the dim object a quarter mile upstream. The crews of the two ships could see the roots of a huge old sycamore tree sticking out of the water looking a little lighter than the dark shadow of the waters. Approaching at the pace of a fast walk, it drifted on a path that would take it between the two ships and intersect the towline that connected them.

Barney broke the silence. "Buddy Joe, I see the sawyer now. It's coming closer to the *Queen* than your boat. God, if that thing gets tangled in the towline, it'll rip the *Queen* off the snag, probably tear our bottom right out in the process. What are we going to do?"

"Don't panic, Barney. Be prepared to cut the towline if it gets tangled with the tree. Maybe we can let the sawyer slip on through and then get another towline reconnected to your boat." Buddy Joe did not think much of that idea but nothing else came to mind.

"Engine room, get the engines warmed and up to speed. We'll have to be moving soon." Buddy Joe prepared his boat for response.

Barney called over the radio. "Buddy Joe, it looks like the sawyer stopped. It's not getting' any closer. Do you think it could have hung on the bottom? It's big enough."

Buddy Joe prayed that luck had been with them. "Jeff, shine your light over on the *Queen*." In the glare he saw the *Bella Queen* drifting away from the bridge.

"Barney, the surge got here before we thought it would. You've slipped off the snag and your boat is making headway down the river. In about two minutes you'll reach the end of the towline and that will swing your boat around. It should pull you out of the way of the snag, but have your crew stand ready to push it off if it gets too near your boat. Those roots can still cause you grief if they brush the *Bella Queen* on the way by."

Buddy Joe instructed Jeff to return to the wheel house and continued to ready his boat for the added weight of the excursion boat at the end of the towline.

When Jeff arrived, Buddy Joe said, "Mate, get this ship into reverse props and lighten the strain on the anchor chains. Just don't run back over them. We're going to get a big yank when the *Queen* reaches the end of that line, and it's going to take both the anchors and the engines to stop and hold her, especially if that sawyer grabs hold of Barney's boat."

A crew man shined the light from the sawyer to the *Queen* and back. It was a race to the end of the towline. Buddy Joe said, "Weve got to pull the *Queen* back across the path of the floating tree to get her clear."

The towboat and barges shuddered as the towline started to draw tight. With a loud twang, the forces pulled the rope clear of the water and it skipped across the surface, straightening to form a direct line between the two ships. All of a sudden the *Bella Queen's* progress down the river halted and it began to rotate and slowly drift out of the main current and toward a position below the *Lady Bird Jamison*.

"Buddy Joe, we're in tow now, but that damned sawyer's coming at us gangbusters. I have the crew on the lower deck with boat hooks and anything else they can find to try to push it off."

Buddy Joe said a large prayer for the excursion boat. This was the most critical time. "Engine room, full reverse." By applying even more tension to the towlines, he hoped to hurry the *Bella Queen* on its sideways journey. He listened to the radio.

"Buddy Joe, it's coming into our port side, about amidships. The boat's swinging around, but it looks like some of the tree's roots are going to claw through the side structure. Oh God, there it goes. It took out the number four and five life rafts, the ones with motors, and I have at least three crewmen down on the deck. Hopefully, no one was pulled overboard."

Buddy Joe could hear heavy breathing and muttering over the radio. He heard Barney yelling something to his crew. Then, "Buddy Joe, it's clear. I think we made it. The snag is clear."

"Barney, get your crew below and check it didn't rip a hole in your hull. And start winching in the towline and get yourself up against the barges. If you have a bad leak, your passengers and crew will have to abandon ship and climb onto the barges. Let everyone know what to expect."

He leaned out the window. "Jeff, go back down and if the *Queen* stays afloat, prepare to tie it onto the side of the barges."

Barney came back on the radio. "Thanks, Buddy Joe. You saved us. I have my crew on that already. Thanks."

"Sure thing, Barney. It's just another day on the river." Buddy Joe blinked his eyes. He turned to engage the controls of the *Lady Bird*. "Engines full stop. Let's see if the anchors will hold both boats."

Buddy Joe sat back down on the stool, leaning forward to place his hands on the railing at the front of the wheel house. He dropped his chin against his chest and sucked a deep breath into his lungs. He felt like he needed something, but he didn't know what. Maybe a nap.

— 13 —
DAWN

Loretta stared around in the growing dawn, her legs wrapped around the spire of the small church, water lapping against her bare breasts. Little remained of the rest of Caruthersville. She could see some treetops and a few telephone poles leaning at odd angles, but of all the buildings in town, only her steeple remained above the waters. What kept it standing? How had she found it in the darkness?

She remembered the flood surge in the middle of the night, the frantic scramble as she and Jim tried to climb high enough to escape the rising waters. They had scaled the walls of the courthouse where they had taken refuge, but the water rose and the building crumbled faster than they could climb. Soon the water engulfed them up to their waists, then to their necks. Loretta remembered holding onto the edge of the broken window frame, floating in the water and letting the building wall protect her from the current.

"I can't swim." Jim had yelled. "Help me." He had reached for Loretta, trying to find something on which to hold, but his fingers slipped away from her bare skin. She recalled his voice as the current swept him down the street. "I can't swim. I can't swim. Help. I can't sw…" She had waited for more, but she only heard the swishing sound of flowing water around her.

That dumb fool couldn't swim? And he was raised alongside the river. Her thoughts dwelt on the incongruity of how someone raised beside the river couldn't swim.

The river rose another foot, then another, until she floated above the building and had to reach down into the water to hold onto the topmost part of the window frame. The current pushed against her body, harder and harder, and her hands tired.

A quarter moon had risen in the east, brightening the surroundings enough for Loretta to see that very few things still stood above the water around her. But about 100 yards downstream she spied a small church belfry.

"You can make it, girl. You can make it." She released her grip on the submerged wall of the courthouse and kicked off, swimming across the current toward the higher edifice. The current pushed swifter than she expected and she almost missed the spire. She swung behind it and had to swim five yards back before she could touch it and grab hold.

The peak of the steeple protruded six feet above the water. The water rose no more, and Loretta felt good about her point of safety. She searched for some way to make holding onto the steeple less tiring, but she found nothing.

That had been five hours ago. Loretta still clung to the belfry. Her body immersed in the water to her shoulders, she trembled from time to time. Her hands cramped, her eyes blurred. Several times she almost fell asleep, but she always roused before losing her grip.

The sun rose above the horizon to the east and glared off the water. Searching desperately, she could see no other living soul. Only she, the spire, and scattered treetops comprised the remnants of the town of Caruthersville.

When another local aftershock shook the ground beneath Caruthersville for half a minute of high intensity shaking, the small church beneath the water under the steeple could no longer stand the pressure of the current and the shaking of its foundations. It swayed to the side and then settled beneath the surface to join its fellow structures littering the flooded streets below.

Loretta cast herself loose, adrift in the muddy waters flowing out of the lost town. She lay back in the water and floated, letting the current carry her where it may.

Her mind drifted and she talked to herself. "I've always been good at floating, that's what mother said." She closed her eyes. The cold water numbed her shoulders, but the sun felt warm on her face. She spread her arms for stabilization as she floated down what had once been

Ward Street. It felt good to be floating. Maybe she could relax and sleep for a few minutes. Maybe she could rest.

In silence the current carried her farther and farther south, across the flat, water-covered farmlands of the Missouri Bootheel.

Chris rolled over, his sore head lying atop a piece of dead wood. The wet grass brushed his face, and he started to sneeze. He opened his eyes to see a line of people huddled along the side of the levee, some still asleep on the ground, others sitting up and holding their knees.

He returned to his back, and then reached down to grab his knees so he could roll up into a sitting position. His arm felt wet and it itched. "Damned chiggers."

Alex and Tina rested beside him, entwined on the ground. "Hope they enjoyed themselves," Chris muttered to himself. "At least the temperature is still in the 70s, but the humidity must be in the 90s."

He stood and brushed dirt off his bare legs. His clothes looked and smelled like the ground. He turned and searched for a bush. Spying one down the levee from where another refugee had just departed, he stumbled that way to find relief. It appeared that life as a refugee would prove to be very uncomfortable.

Returning five minutes later he reached down and shook the couple by an exposed shoulder. "Hey, Alex, wake up. If you don't move some muscles soon you'll freeze in that position."

Tina opened one eye and looked out the side to Chris. "What do you want?"

"It's going to be hot, and if we're going to Memphis we should leave soon. Besides, we may have trouble finding food, so we need to hurry."

Alex sat up and pulled Tina up beside him. "My bones hurt. Sleeping on bare ground is the pits."

Chris laughed with the optimistic sarcasm of youth. "Get used to it. That's our life until who knows when. That, and doing without food."

Buddy Joe checked his wristwatch against the ship's clock, which was soon to ring four bells: Sunday morning, five minutes to six. The rising sun pulled the shadow of the I-255 Bridge back across the river from the western shore. Just outside the wheel house door on the port side,

Jeff and Virgil talked while pointing to the shattered granary just below the bridge, but he could not catch the gist of their conversation.

Buddy Joe leaned back against the starboard window in the wheel house of the *Lady Bird* and watched Barney connect the small TV to the cable running through the window. He had gotten four hours sleep once he felt satisfied that the crew had tied the *Bella Queen* correctly to the barges. He needed the rest, and he wanted time for things to settle before moving. This mooring had proved to be the safest harbor for the time being.

He leaned forward to scan across his tow of barges, now enhanced by the excursion boat. "You know, Barney, in the light I can see the river is running higher than yesterday but not as high as I expected. It looks like it's just barely covering the pier where Virgil fished."

Barney looked out the window. "I see what you mean. That surge must have been just that last night, up then back down. If it hadn't kicked us loose from that snag, the *Queen* would still be sitting there under the bridge superstructure."

Buddy Joe rubbed the growing stubble on his chin. "My guess is that a lot of water must be flowing out of the river onto the surrounding land. As time goes on, we may even see the river drop some more."

"Flooding all that farm land could soak up a lot of water. That might be good for us but hell for the farmers."

Barney adjusted the controls on the small TV set and exclaimed. "There, got it. This is a direct feed from our satellite TV receiver in the lounge. I told the purser to leave it set to the Network Worldwide Satellite News program. They seem to have the best coverage so far. Thought you and your people should hear their 6 AM summary."

Buddy Joe took another sip of his black coffee. "Thanks, Barney. Maybe with some daylight and information we can start to make sense out of this."

Carolyn Phelps had never left the studios of NWS. She had slept on the couch in the women's room. She felt gritty, but there was no one else available. The red light appeared on the camera and she smiled mechanically, live on the air.

"It has been almost 21 hours since the monster earthquake struck the Mississippi Valley between St. Louis and Memphis. Satellite imagery and airborne cameras are beginning to show us some of the devastated areas." The screen showed aerial scenes of the Mississippi River flooding the surrounding lands.

"We still only have sparse reports from the most damaged regions. Most roads into the areas along the Mississippi are mostly flooded, and the police and the Army have restricted access into the region. Telephone service was lost during the shaking; only a few contacts by two-way radios have been made."

Carolyn Phelps looked into the camera from the NWS news desk. She knew the makeup crew had done a good job covering the haggard lines that must course across her face. The director lifted his palms, telling her to put some verve into her delivery. She smiled, machine-like.

"Yesterday, President Frank LaPorte declared Martial Law for the Ohio and Mississippi River valleys and we are told that Army units began to arrive in the Memphis area last night.

"The United States Geological Survey now pinpoints the epicenter for the magnitude 7.9 earthquake as beneath the little town of New Simon, Arkansas, at the state boundary in the bottom of the Missouri Bootheel.

"A second earthquake with an epicenter southwest of Tiptonville occurred less than two hours later, registering a magnitude of 7.4. Numerous strong aftershocks, two as high as magnitude 7.0, continue to pound the region, causing even more damage as far away as St. Louis, Cincinnati, Nashville, and Little Rock.

"Aerial reconnaissance shows that fires still rage around Memphis and on President's Island in the industrial complex just to the south of the city. All bridges connecting Tennessee with Arkansas are impassable, though there appears to be foot traffic crossing on a railroad bridge between West Memphis and Memphis." The TV screen showed a view of the city of Memphis. A smudge of smoke covered much of the city, but some individual structures could be picked out. The Pyramid lay in a heap near the river.

"Flooding is becoming the major concern throughout the stricken region." The view switched to a plume of water gushing out the side of a monstrous earthen dam. "Water continues to flow through the broken locks at the Kentucky Dam, releasing the contents of the lake into the

Ohio River just above Paducah. Authorities estimate the flow has dropped to just under 50,000,000 gallons per second and say it will continue for days. Flooding below the dam is widespread. Thirty percent of Paducah is now under water and flooding has spread across southern Illinois and into the Mississippi River valley."

An aerial view of the New Madrid bend filled the screen. "The Army Corps of Engineers confirm that the Mississippi River has cut a new channel near the old river town of Bessie on the New Madrid Bend, creating a mile-long rapids that replaces the 19 miles of river around the bend. Several boats have been sucked into the waterfalls, and the fate of their crews is unknown at this time."

"Along the Mississippi south of Memphis, communities are busily repairing levee breaches along the river. At this point it appears that the massive structure separating the Mississippi River from the Achafalaya River has been pushed aside, and a flood has swept down its channel to the Gulf of Mexico. This flood has destroyed Morgan City and other low-lying towns in Louisiana. The Army Corps of Engineers are not commenting on what can be done to stop that flood.

"In New Orleans, crews are struggling to repair the many broken levees that have allowed the river to flood all low-lying areas of the city. The surge from the waters of Kentucky Lake will not reach that portion of the river for another three days, and authorities hope to have enough repairs completed to prevent even more widespread flooding of the city.

The camera returned to show Carolyn, her eyes now tearing and smearing her makeup. She used a napkin to blow her nose. "Pardon me."

Swallowing, she continued, "Elsewhere, reports of fatalities and injuries continue to come in. Deaths are reported in Chicago, Indianapolis, Cincinnati, Nashville, and Atlanta. Authorities in St. Louis now estimate there have been over 1,000 fatalities in that city alone, with injuries approaching 12,000.

"Massive casualties occurred in and around the parking structures next to the Arch where a large celebration was underway and at the site where the old Union Station collapsed onto a Science Fiction Convention. We now have confirmation that the President's daughter was one of the casualties at Union Station.

"Amazingly, the Arch withstood the shaking of the earthquake, though police say that over 300 people were trampled to death from the shaking and panic in the museum below the arch.

"Casualties throughout the rest of the country are still being tabulated."

Carolyn dropped her eyes, unwilling to continue facing the camera. Her shoulders showed her level of tension. Her voice broke as she said, "And now this from our sponsors."

Buddy Joe walked out onto the bridge deck, shaken by the news report he had seen on TV. The trees along the banks still stood and the water still flowed, but he felt a sense of dread. "Barney, we're looking at the end of our world, at the end of the river and everything that ever was normal here." His hands gripped the railing and he squeezed hard, trying to ease his tension.

His mind clouded over and he felt unsteady for a moment. The image of a racecar tearing through the pits again flashed through his mind, and then disappeared. The image of his barges bucking up and down, straining at their lines, replaced it and filled his mind. The river reached out like a monster to tear his boat and crew apart. "Thank God, Jimmy doesn't have to see this."

He shook his head, trying to clear his eyes. "Barney, we've got to find some way to keep from going insane. We've got to go totally into survival mode. We've got to save whatever we can of the world we knew."

Barney leaned over the rail and looked ahead at the covered barges, now home to some of the hardier refugees rescued from Caruthersville. The *Bella Queen* was filled with those less able to withstand the outdoor elements. "I know what you mean. What are your barges carrying this time?" he asked.

Puzzled, Buddy Joe looked at Barney. "Soybeans and corn. Why?"

"It's like you said, Buddy Joe. We're in survival mode now. You're not carrying a load of grain; you're carrying a load of food. Thank God it's not a load of rock. Hard to eat rock, even when you barbeque it."

Buddy Joe thought a moment then felt a pressure lift from his shoulders. His friend's joking had shown him a way. Turning to Jeff, he laughed and said, "Mate, go take this comedian to recheck that the

Queen is tied in for the tow. And make sure he doesn't start eating our cargo, at least not yet. We'll head downriver in about an hour."

Buddy Joe turned when he heard Virgil say, "Cap'n, there's a loose barge coming down the river, and there's somebody on it."

Buddy Joe reached back into the wheel house for his binoculars and scanned the river to the northeast. He could see the barge half a mile upriver. As he brought the scene into better focus, he saw a young man standing on the cover at the front of the barge. Then, what looked like a young woman holding a baby climbed out of the tool shack to join him.

"Cap'n, I see two people on board. Can you tell anything about the barge and where it came from?"

"Virgil, you've got eyes like an eagle. Looks like one of those old secondhand barges tied up at Cairo. They collect them there to pull out and scrap. I would guess that the couple on board must have taken refuge there when the flood surge hit."

On the drifting barge, Freddy steadied Paula as their craft drifted toward the bridge. "Look, Paula, there's a couple of big boats. They can help us." The two waved their arms, trying to attract the attention of those aboard the towboat and riverboat.

"Help, help," they screamed at the top of their lungs as they flailed their arms with frenzy. Freddy pulled off his shirt and waved it up and down.

The barge turned in the current as it neared the bridge. Freddy said, "Paula, they're getting ready to save us. See the big man on the wheel house deck with the bullhorn? What's he saying? Do you understand him, Paula?"

The voice came across the waters. "… out. Get down. You … snag." The big man waved his hand up and down.

Paula said, "I heard him say something about a snag. Do you think something is wrong?"

The barge drifted sideways under the bridge. The pair of refugees stood on its bow, the end away from the towboat. Freddy scanned the river. "Snag?" Then he spied the dimple in the water. "Down, Paula, get down. That's what he said." Freddy grabbed Paula holding the baby and pulled her to the small deck space.

As the stern of the barge passed under the bridge, the fallen road section once again gaffed a victim. A loud screech of metal tearing metal pierced the air. The barge shuddered and whipped around its stern, now held in place by the snag, throwing Freddy and Paula against the tool shed. With his free hand, Freddy grabbed the door and held on. With his other hand he clutched Paula and saved them from sliding over the side of the barge into the water.

Another screech sounded as the barge pulled free of the snag and settled in the water to continue its course down the river.

Buddy Joe's amplified voice reached out across the water. "Hello, are you okay? Wave your right arm if the answer is yes, left arm if no."

Freddy stood and waved. "Yes, yes. We're okay, but we need someone to get us off this barge. Help us, please." Paula stood and together the couple waved their right arms up and down and hugged each other.

"I see you're okay for now. We're not underway yet so we can't help you. It looks like the barge survived the collision with the snag, but it could be taking on some water. I'll radio downriver that you need assistance, and if we see you again we'll do our best to help."

The couple dropped their arms to their sides and their shoulders slumped. "Freddy, why can't they throw us a line? Don't they have a boat to come and get us? Who is there that can help us downriver?"

"I don't know the answers, Paula. I don't know." He watched as the barge passed within a 100 yards of the towboat and continued downriver. The towboat remained in sight for 15 minutes, never moving.

Buddy Joe thumbed the microphone. "Calling the *Amanda Blair*. Charlie Green, are you awake yet?" He waited a moment then tried again, "Charlie Green, wake up."

He had almost decided to give up when a voice came back over the radio. "This is Charlie Green of the *Amanda Blair*. Come back."

"Charlie, this is Buddy Joe of the *Lady Bird Jamison* up at the Boothspoint Bridge. A loose barge just passed us floating high like an empty. It bounced off the snag under the bridge. I think it's okay, but it could be taking on water. There's a young couple on board, and it looks like they have a baby with them. They need help. Think you can get them off the barge?"

"We can sure keep an eye out for them. Sylvie's pleased as punch that she's got a couple of refugees to take care of now. The more the merrier." He paused and then asked, "Captain, we're anchored just up from Ashport. How long do you think it will take for the barge to get here?"

Buddy Joe looked at the map. "You're about 35 miles below us, and assuming the barge stays in the main current I figure it will be there in a seven to ten hours. You better come upriver just a bit to meet it in a straight stretch. You don't want the currents in the bend to wrap it around you."

"Very good, sir, and we will hold in that area to meet up with you."

Buddy Joe responded, "Okay by me, but I expect to anchor tonight around Barfield, so it'll be tomorrow sometime before we reach where you'll be anchored. We'll be another hour getting underway. I want to make sure we have everything tied down before I move my tow very far. We'll keep an eye out for you. *Lady Bird Jamison* clear." Buddy Joe released the button.

Virgil grinned at Buddy Joe. "Yo' shore are getting into this refugee business. Must be that God has that picked out for you, just like he has picked me to be his disciple. He baptized me in the river, you know."

Buddy Joe grinned back, recalling Virgil's tale of his baptism. Then, after thinking a moment, he said, "Maybe you're right. Maybe that's what I can do, go into the refugee business. Looks like there'll be some demand for that kind of job for some time to come."

— 14 —
MOVING DOWNRIVER

"Cap'n, the *Bella Queen* is secure and the barge ties are all retightened. We can move downstream at any time." Buddy Joe sat up and accepted the cup of coffee from his first mate and took a sip. It burned his tongue. "That's too hot. You forgot the ice cube."

"Sorry, sir. I like it hot, and besides, we're out of ice."

"You're just like Jimmy. He always fixed it too hot. Thanks anyway, Jeff. Let's go forward and have one last talk with Barney before we move out." The two walked down the ladder to the deck of the *Lady Bird* then across the barges toward the excursion boat. Several refugees stepped forward to shake their hands. Barney Ruggs met them on the deck of the barge next to his boat.

The *Bella Queen*, its stern pointed in the direction of travel, had been secured to the side of the first set of barges. From bow to stern ropes that had been strung to the corners of the outer barges, were pulled taut to hold the riverboat steady.

As the three looked over the ties holding the excursion boat to the barges, Buddy Joe reiterated to Barney how they would proceed. "The *Bella Queen* is now a *de facto* barge and the *Lady Bird* will control all its motions. Barney, your primary responsibility is to see that your passengers remain under control and out of danger."

"Can they help?"

"They can act as lookouts. You stay around your pilothouse, near the ship-to-ship radio. Use your handheld radios to talk to the lookouts and relay the information to me. Your boat blocks some of my view, but not all. I can still use some help up front. I'll have members of my crew up there with them part of the time."

"Sure thing. I think some will like that kind of duty."

"Barney, you might explain to your passengers a towboat has to oversteer so that it can bring the long line of barges around to the proper heading to proceed downriver. From time to time they may feel I'm steering them into the bank, but that is not the case. They just need to have patience and a little faith in the Captain of the towboat."

"Don't worry. They all think you're their savior."

"Knock it off, Barney. Anyway, I'm going to take it easy and hold to the middle of the river. Tell your passengers that we'll anchor at night, so it'll be a couple more days on the river before we reach Memphis. Think your folks can hold out that long?"

"Sure thing, Buddy Joe. We'll just sit back here on the *Queen* and take life easy for a change. If you get bored come forward and we'll serve you a mint julep." Barney smiled as he shook the hand of his fellow river man.

"Come on Jeff, let's get back to our job before this man convinces us to give up our hard life of honest work on the river."

An old man in overalls stood and pointed. "Look, just below the bridge, that towboat's getting underway. I sure was happy to see they had that excursion boat tied up next to its barges this morning. Looks like they'll be able to make it down to Memphis."

The woman sitting on the ground next to him shook her head. "I don't see why. Memphis must look worse than Dyersburg, and our town has been destroyed. There ain't nothing' left of our place. What are we going to do, Thomas?"

"Pray, Nelda. For now you just pray." He nodded to the young girl walking lamely by with two young men back toward Dyersburg.

Tina stopped and asked, "Do you know how far it is to Memphis from here?"

"It's about 80 miles from Dyersburg to downtown Memphis," Thomas said.

"Thanks. Guess we should be able to walk that in about four days."

The old man looked surprised. "Ya'll going to walk all the way to Memphis. What in the world for?"

"My dad and Chris's dad are there, and we have friends who are there. Maybe we can help them. Besides, what else is there to do?"

The old woman looked up. "You can pray. That's what Thomas said to do, but I don't know what to pray for."

"Now just hold the rod with your right thumb on the spool to keep it from rolling out. Hold the rod level and steady with your left hand and sort of bend the line with your fingers up in front of the reel. Keep the line just straight, not too tight. You want to be able to feel if a catfish starts to nibble on your bait."

Charlie Green instructed Samantha on the fine art of river fishing as they sat on the *Amanda Blair's* fantail deck 50 yards off Daniels Point Landing.

Across the river he could see three fires still smoking in Ashport, but he had already talked with the survivors there and they declined his offer to help.

While Charlie waited for the barge to appear, he tried to keep his young charge occupied. In the meantime, Sylvie told Danielle all about her own grandchildren while she redressed the girl's snakebite.

"Uncle Charlie, I feel something, like it's hitting my line."

Charlie jumped up, eager to help Samantha. He watched the tip of the rod bouncing up and down a half an inch. "There. That's a catfish nibble. Just hold it steady. Wait till he grabs it in his mouth and starts to pull. That's when you hook him."

Charlie pranced around the young girl, all thoughts of the earthquake pushed to the back of his mind by the immediacy of catching a fish.

'It feels like he's pulling harder." The tip of the rod bent down, an inch, then two.

"Now. He's got the bait in his mouth. Quick, lift the whole rod up and hard back. There, you've got him. I can see him fighting. Oh boy, he's a big one."

Samantha squealed with joy.

Passengers and refugees on the excursion boat lined the railings, watching as the towboat pulled its flotilla back toward the bridge and lifted its anchors. At last the procession moved out into the main channel of the river and the faster current.

The expected flood surge had arrived and passed. In the middle of the river one could hardly tell that the river still ran above flood stage. From the second deck at the stern, Ron, Tim, and Clarence stood ready to radio any unusual sightings to Captain Ruggs.

Ron said, "Okay guys, we need to watch the river where we're going, especially for big chunks of debris. I talked to one of the towboat crewmen last night. He said we would do about three knots in this current. He said the towboat would not be pushing so much as pulling to keep us from going too fast. There's us and another lookout up on deck three watching for snags or any other problems."

Clarence asked, "How does Captain Simpson know where to steer the barges?"

"The crewman told me they normally have buoys marking the channel in the river, but most of them are gone now, both from the flooding and the earthquake. I'll tell you though, this is almost as bad as flying over the Saudi desert. There just aren't any landmarks. However, they have a backup GPS with precision maps, so the Captain knows where the river is supposed to be."

Water covered the flood plain of the river and reached through the trees to a thin line of land that Ron concluded must be what remained of the levees. Tim pointed out an obvious break in the levee to the west. "Right there, it looks like the levee is missing. At least the soil and tree line are missing. What do you think?"

Ron's practiced eye from his pilot days did not fail him, and he agreed. "You're right. There sure seems to be a lot of water flowing through that break in the levee. It looks like the fields on the other side are completely covered with water."

Clarence looked at the map and shook his head in dismay. "That would be about where Cottonwood Point is located. If the river rose seven feet like they expected and is flowing that hard out the break, Cooter, Steele, and New Simon must be under five feet of water by now. Oh God, what a mess."

Ron pointed east. "Look, on the other side. It looks like there must have been a small community there. I can see some housetops and telephone poles."

"That should be Heloise," Clarence said. "I have an aunt who lives there, or at least she used to."

They saw little to report, though the occasional propane tank floating down the river a bit faster than the boats got everyone's attention.

Three miles farther along, Tim pointed downstream to where the Obion River joined the Mississippi from the east. Again, he saw another break in the levee and knew there was more flooding. "Oh, that's terrible."

Clarence shook his head. "That's where we came from."

"Come on, guys, quit thinking about it. Just relax and enjoy the trip down the river," Ron suggested.

Apart from the drone of the diesel engines on the towboat a 100 yards to the rear, peace enveloped the river and little suggested that the land around them lay gripped in the aftermath of a giant earthquake. The water no longer acted troubled and chaotic, and its muddy color could be due to normal springtime floods from upstream.

Soon, many of the crowd along the rails moved out of the sun to cooler places inside where they could play games or do whatever people do on an excursion boat when it is underway.

"Uncle Charlie, why are those people climbing into the trees across the river?"

Charlie Green roused himself from his daydreams and picked up his binoculars. "You're right, Sam, they're climbing into the trees." He carefully inspected the shoreline and then looked at the near shore behind them. "I think there's a problem."

Charlie moved forward to the bridge and picked up the radio microphone. "Charlie Green calling the *Lady Bird Jamison.* Captain Simpson, are you there?"

Buddy Joe reached over and took the handheld microphone. "This is Captain Simpson of the *Lady Bird Jamison.* Hi there, Charlie. We're underway now, and everything looks clear. We're just past Island Twenty-One approaching the Obion River. What's up? Where are you located? Have you seen the barge? Come back."

"Captain, we're still just across from Ashport. I'm looking for the barge but haven't seen it yet. I was letting the girls we rescued do some fishing to pass the time. But now I'm seeing a problem. The river seems to be rising and rising almighty fast. I don't know why. Twenty minutes ago there was some land showing in front of Ashport, but now it's gone and water has spread around the few houses still there. There were some people standing around, but they've started climbing up into the trees and onto the roofs. Understand?"

Buddy Joe thought a moment then responded. "The surge from Kentucky Dam has already passed you, so that shouldn't be what's raising the river."

Charlie continued. "Right, it must be something else. I was thinking. Remember that earth slide down around Fort Pillow? Well, I think it may have dammed the river, and what we're seeing is the water backing up behind that dam. Is that possible?"

Buddy Joe scratched his head. "I suppose that could be happening, but it would have to be an awfully big dam to raise the water level that far upriver." He studied the map. "Ashport is 15 miles above Fort Pillow, but the topographic map does show there's less than a 10-foot difference in river elevation, so I guess it could back up that far."

"What will that do to the flooding around Osceola?"

"It would make a huge difference, and if it's a dam I'm wondering how I'll get this tow around it when we get down there."

"What should I do now?"

"I'd suggest you keep an eye on Ashport and figure out how high the river is getting. You may need to do some rescuing if the water is climbing like you say. This afternoon you can motor upriver to around Tomato. That's about as far as I want to go today. We need to put our heads together and think this out. And keep your eyes peeled for that barge. Hopefully, it hasn't already sunk or found a crevasse in the levees and been sucked out of the river."

The barge spun in a very slow circle as two currents fought for it's possession, one flowing south down the river channel and the other flowing west out the huge levee break north of Barfield. After what seemed to be an hour, the river won and the current sucked the barge back into the main channel to continue its journey down the river.

"Where are we, Freddy?"

"I don't know, Paula. There're some pieces of big buildings still standing west of us. It looks like it could be a heavy construction plant, so at least we're in civilization. But I haven't seen anybody alongside the river for a long time."

"I'm hungry. Did you find anything else to eat?"

"No, there was that little bit of stale wheat in the corner of the barge, but that was all. I haven't figured out how to fish yet."

"Can I have another drink of water? The baby must be sucking everything out of me, I feel so dry. My lips are chapped. At least she seems to be happy."

"I brought up another pail of water a while ago and set it in the shade. Some of the silt has settled, so at least you don't have to drink mud. Maybe it'll rain again and I can get some clean water."

"Thank you, Freddy. You do so much to take care of us. I love you."

Freddy scanned the near shore, hoping to find some help, but he saw none. He couldn't help the tears that filled his eyes as he looked away. "I love you, too, honey. Both you and our baby."

Jud and Sally Mae rode the stolen jet skis through the vacant streets of Osceola, staring at the damage left by the earthquake and seiche, now stirred again by the rising water.

Jud pointed. "Hey, look there. That's Jake's Bait Shop. It shore don't look like he'll be buying anything more from me for a long time."

He drove his jet ski up to the door of the building. The water stood three feet deep at the front door. He reached over and shook the door. It bounced open a crack.

The building on the south side of Jake's shop had been crushed into the concrete wall separating the two buildings, as if it had been hit by a tornado. The front windows of Jake's shop had all been knocked to the inside, and jagged glass lined the frames. The metal-frame roof had fallen and tilted to the back of the store.

Jud looked back and grinned. "Sally Mae, pull the booty boat over here. Let's see what we can take from old Jake. That bastard never did pay me enough for what I brought him. He owes me."

Sally Mae's jet ski pulled the small boat in which they collected their loot closer to the door.

"What if he shows up, Jud? Don't you think he might get upset?"

"He won't be coming around. Ever' body has headed for higher ground, what with the water rising the way it is." Jud stepped off his jet ski into the waist-deep water and tied the anchor rope to a bar in the window. "I'll just take a look." He pulled the door open and peered inside the dark room. "Looks like there's been quite a bit of damage with the roof falling in and things falling off the wall."

"Jud, check if he has any beer. And maybe some cans of food?"

Jud stepped into the building with care, feeling with his tennis shoes across the floor under the water. A small aftershock rattled the ground and he froze, not used to feeling the shaking of the earth. "Oh God. That don't feel so good." Most of his time had been on the water.

After a few seconds the shaking went away, and Jud continued his inspection of the building. He waded to the refrigerator and with considerable effort opened it. Water poured into the cavity, filling it halfway, but not before he saw cases of Coors on the first shelf. "I found you some beer," he called. He hoisted two cases and waded back to the door to place the loot in the small boat.

Bending over to avoid bumping into the fallen ceiling panels, he waded to the cash register at the back of the store. He tried to find someway to open it. His toe felt a soft object under the water, and he pushed it away with his foot.

Reaching under the cash register to search for keys, he closed his fingers around a metal object. "Hey, Sally Mae, I found Jake's pistol."

He brought the chrome-plated automatic to the surface and draining water out of the barrel, placed it on the counter. Reaching again, he found a sodden box of shells. As he tried to pick it up he felt it collapse, and he grabbed a handful of shells.

"What do we need with a gun?" Sally Mae asked.

"You never know. It might come in handy." He stuffed the shells into his jeans pocket.

Jud bent down to reach deeper into the shelf. From the corner of his eye he saw a white bloated object rise to the surface of the muddy water. Blank eyes stared through the murky fluid at Jud. "What's that?" he cried as he jerked back. Grabbing the pistol he fired a shot into the apparition in the water beside him.

The water exploded from the impact of the shell and gases, drenching Jud who had pulled back against the wall.

Then he realized what he saw. "That's Jake. What's he doing here? He's just floating in the water." The body rolled over.

"He's dead." Jud turned and ran through the water as fast as he could, outside to his watercraft. "That's Jud in there. Let's get out of here. Quick."

Turning the jet skis back to the river, the pair sped out of town with only two cases of beer, a pistol, and a pocketful of shells.

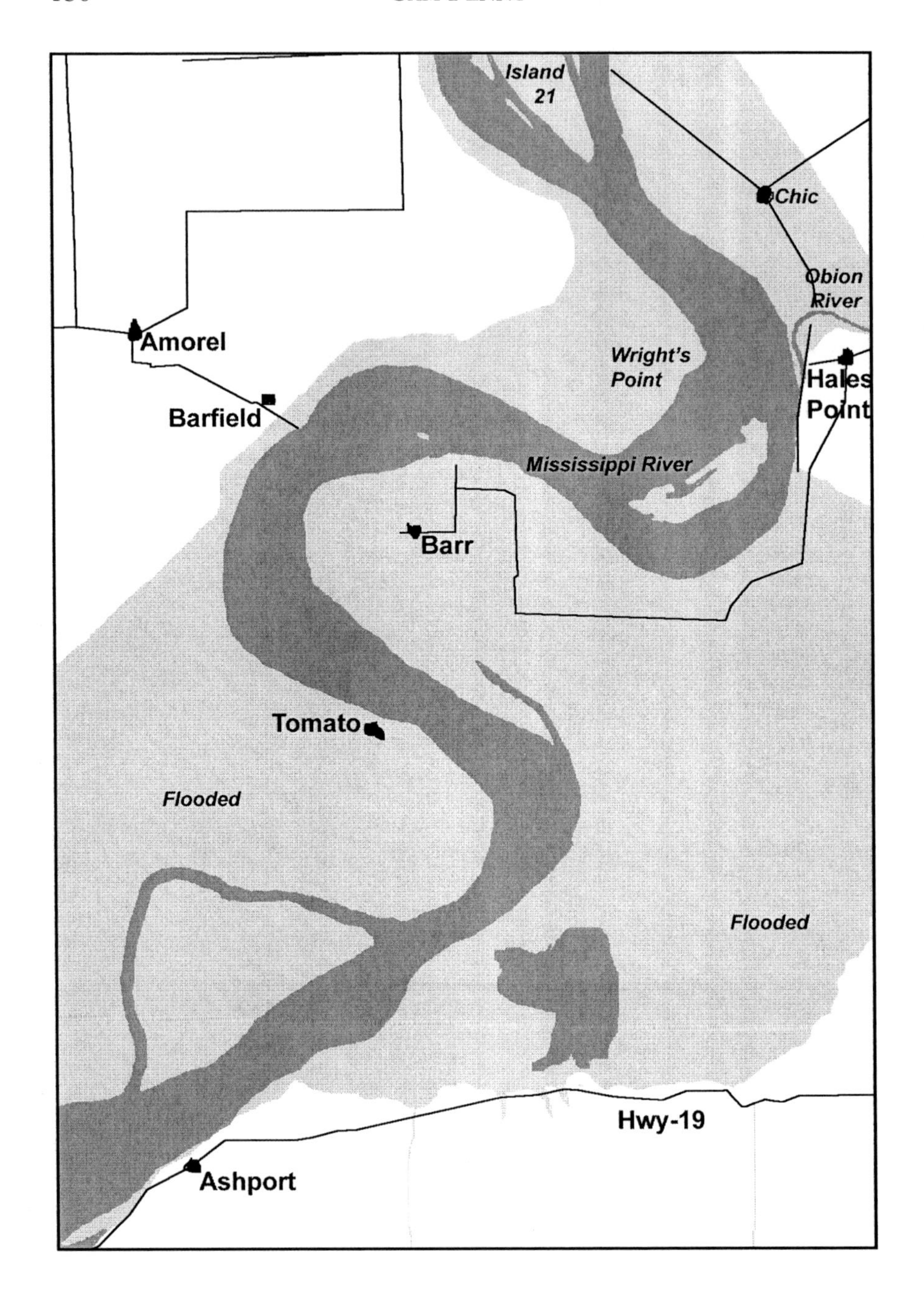

Barfield Levee

— 15 —
Fighting Back

"That's where our house was, at the end of Route 21 next to the Obion River. Half a mile up that road is what's left of the little store at Hales Point." Clarence pointed to where a straight road cut through the trees heading eastward. Ron could see water covering the roadway for at least a mile away from what had once been the bank of the river.

Clarence continued, "The Obion River comes into the Mississippi along here. It was flooded with Mississippi River water, but we still had a couple of feet to go before it got to our house, and the house was on stilts, so I wasn't worried. My wife Suzie was, but I wasn't.

"Then when the earthquake hit, the house fell off its stilts. For 100 yards back from the riverbank the ground dropped a good five feet, so we and our house was under water. Lucky I had my boat sitting loose on the trailer in the front yard."

Ron asked, "So what did you do?"

"Well, after the earthquake, my truck was under four feet of water so I couldn't get it started, but I untied the boat from the trailer and loaded Suzie and whatever we could get out of the house into the boat. I paddled up the road to the Smith's place. Their house had fallen, too, and it was burning, lying there in two feet of water. Jesse, their kid—he's about 10—was outside crying. He said his folks were trapped inside and had died in the fire. It smelled awful, but there wasn't anything we could do.

"We tied the boat to a bush and took Jesse and walked up to the little store at Hale's Point. Old Missus Jaques was pulling away at the pile of lumber, trying to get to her grocery goods. When she found a can or box she'd set it out beside the road. When we showed up she yelled at us, said everything was double price from now on and how she wouldn't take no checks."

Ron shook his head. "It's funny how some people react in times like this. So how did you hear about the surge from Kentucky Dam?"

"We went back to the boat and paddled back to the house. I had my marine radio on the dash of the truck, so it still worked, but there weren't many people talking. But that was when I heard the riverboat Captains talking about the break at the dam and how it was going to raise the level of the river lots.

"I told Suzie we should find some higher ground and more civilization. I knew the road would be flooded back to the east, so I started up the boat motor and we headed up the Mississippi, hoping to get to Caruthersville.

"We were about to sleep in the boat alongside some trees when we heard a yell and found the two men who said they were from the *Bella Queen*. We took them aboard and that would a sunk the boat if one of us rolled in our sleep. So we continued up the west side of the river to where I saw the lights of the towboat.

"It sure was nice of the Captain to let us come aboard last night. We were out of water and food."

Ron smiled. "We sure are collecting a motley crew of survivors on this ship. Welcome aboard."

The sun moved past the zenith into the western sky as Charlie steered the *Amanda Blair* past what remained of Lower Forked Deer Landing. He had still not seen any sign of the errant barge, but he watched with care. It could be pushed up against the bank or lodged in some water-covered trees anywhere along the river.

"Do you think those three men will be okay back at Ashport?" asked Sylvie.

Charlie squinted his eyes, looking to the north. "They wanted to stay, even with the water coming up. It's their call. Guess they think it's their job to watch over the granary, maybe because it's split."

Sylvie looked disturbed. "They said they talked with someone by radio, and there was concern about looting. There had been some kind of problem down around Osceola and the government had to send the Army into Memphis. Do you think that could be a problem for us?"

Charlie rubbed his wife's shoulder. "Look, you shouldn't get yourself riled up about such things. Yes, I expect it's probably getting kind of lawless in some areas, but here on the river we're protected from people like that.

"Why don't you go play some more games with Samantha and Danielle? You get so much pleasure out of having them with us you sometimes look like you're glowing."

Sylvie smiled and blushed. "Oh, it's so nice to have children around again. And they are so sweet and courteous, just like I always wanted our grandchildren to be."

Then her face clouded. "But I'm worried. We might have problems with looters. I'm going to get out that shotgun you used for skeet shooting and see that it's loaded."

Charlie sighed. "Whatever you want, dear. Just be careful with it. And you might give the girls some lessons on gun safety, like where the safety latch is and when it's on. For all you know, they may not have been taught those kinds of things."

"You're right. It would be just like their mother to not teach them the right lessons."

Buddy Joe selected the local frequency he and Barney were using and thumbed the microphone. "Barney, this is the strangest I've ever seen on the river coming around Wright's Point. The current's awfully strong through the chute, even with the flood plain covered like it is. The river drops about four feet through here and there's normally sort of a small rapids on the west side of the island, but the GPS shows my speed is almost 10 knots. I'm pulling back at about four knots; so that says the current must be 14.

"Something is sucking the hell out of the river downstream to create that kind of current. What do you think?"

Barney responded after a moment. "Buddy Joe, you said your friend Charlie told you the river was rising down around Ashport. If it's rising there, 11 miles downstream, and running like hell here, somewhere in between there must be some big break in the levee that's taking a lot of water out of the river."

"That's what I'm thinking. Even though we've still got lots of daylight left, I think we should anchor soon and check this out. I'd rather not take this flotilla by some big crevasse without being able to see everything and know what's going on. I wanted to meet up with Charlie around Tomato, but he can come up our way instead."

"Maybe he'll be able to spot where the water leak is on the way."

The first barge rounded the point and Buddy Joe steered the towboat hard to the starboard to bring the back of the tow around to port so it would point downriver. Soon he had a clear view downriver and the tops of some fallen buildings appeared above the trees. "Barney, I'd say the steel mills at Armorel took a real pounding. I remember seeing a whole line of buildings down there. Now there are just those two ends sticking up."

Barney came back. "I agree, and I think I see a dimple in the horizon just upstream from where the Barfield elevators used to be. That may be where the water's going."

Buddy Joe responded, "Okay, I see what you're talking about and I've seen enough. I'm moving this rig to the port side of the river and I'll drop anchor on the inside of the big bend down by Barfield. That's out of the channel and there should be a lot less current in the shallower water. Tell your lookouts to be sharp and let me know if we're running into any debris. There's no telling what's floating in the river now."

"Roger that, Buddy Joe. I'll get my crew busy. And once you're settled in, you and your crew are invited over to the *Queen* for supper."

Buddy Joe laughed. "Thanks, Barney, but me and the crew will stick here on the *Lady Bird* for our meals. You're feeding refugee rations to a bunch of rowdies, and I've got a freezer half-filled with rib-eye steaks and catfish. Maybe we'll come over later."

He continued, "Besides, I want to catch what's happening on the TV."

NWS returned to its mid-afternoon discussion show about the US economy. Harold Owens spoke into the TV camera, "Sorry for that interruption, folks. But now we will continue with our discussion of the state of the nation.

"We are joined by Senator Wright and Congressman Holman via a satellite feed from Washington, DC."

Harold pushed a button and spoke. "Senator Wright, you were about to give your assessment of the impact of this earthquake on the United States."

The Senator cleared his throat and leaned back into the comfortable chair. "Yes, Harold, I just want to say that we are faced with the most challenging time our country has ever seen. There has been major damage in 11 of our states, and significant damage in another 20. Ninety million people have been affected directly by these quakes, and 10,000,000 people are without shelter, food, water, or power at this time.

"Our first concern must to be to come to the aid of those in the most devastated areas. We know the city of Memphis is totally destroyed along with hundreds of small towns in Arkansas, Missouri, Illinois, Indiana, Kentucky, Louisiana, Tennessee, and Mississippi. Other major cities with widespread damage include Little Rock, St. Louis, Paducah, Evansville, and Nashville."

"I commend the President for declaring a state of emergency and placing those areas under full military control. Only the military has the facilities and equipment to get aid to the stricken people in an immediate fashion."

Congressman Holman fidgeted and then objected. "We can't have the military taking over. Consider the cost to the federal government for such an operation. The state militias should be allowed to handle these problems themselves. The President has no right to send in the Army like this." He stared belligerently at the senator across the table. Already, disagreement threatened to get in the way of accomplishing recovery.

"Congressman," Senator Wright held up his hand. "I am afraid that we are well beyond the point where we can leave it to the states. Now you are from one of the states with only moderate damage, and …"

Holman bristled. "Just because we didn't suffer as much damage as Arkansas and Tennessee and Missouri doesn't mean that we don't care about states' rights and states' responsibilities. What I mean is, every state should take care of its own problems. If the federal government takes care of this then everyone in the country will have to pay."

"Congressman, yes, everyone in the country will have to pay. This is not an issue that can or should be solved at the state level. We are talking about the entire United States economy. We are all in this boat together. Everyone in the country has lost a friend or loved one in this disaster. Everyone in the country is affected in some major way by this tragedy. Everyone will now have to do their best to join in the effort to save the center of our land.

"Realize, from what we know, the extent of the damage in the stricken areas is so severe that there is a danger that all those who can will simply pack up and leave, and the ones who can are those with skills that can be used elsewhere. Whole industries will close down, if they are not already closed by the damage. Towns that used to thrive will become ghost towns. The entire Mississippi valley could become a wasteland. We simply cannot allow that to happen. This situation calls for extraordinary measures."

Congressman Holman set his jaw. "If the President continues with this course of action, I believe the situation calls for impeachment. He should let the states do it. It is their problem."

Five miles below the *Lady Bird Jamison* and her tow and two miles below Tomato, Charlie Green held to the middle of the channel and steered the *Amanda Blair* toward Barfield Point. He continued to scan everywhere for the barge with the refugees Captain Simpson had described. He had seen several barges, most of them along the banks, some anchored and a couple loose, but they had all been empty.

Sylvie sat on the back deck playing another silly card game with the girls. Behind them were two fishing rods with lines trolling behind the boat. He had told the girls, "You probably won't ever catch anything on this river, but if you want to troll you can." They had taken to fishing and life on the boat with enthusiasm and seemed to be enjoying themselves to the limit.

Charlie became aware of a high-pitched whine, like a small engine. He scanned the banks for the source of the noise and then looked to the rear. "Sylvie, there's some jet skiers coming up behind us. Looks like one of them is pulling a small boat behind."

Sylvie stood and shaded her eyes. Picking up the binoculars she studied them with more care. The two girls stood and waved.

Charlie watched with interest as they approached. They seemed to be angling toward the boat. He called to the girls, "Better reel your lines in so they don't get tangled with these jet skiers if they get too close."

Sylvie stepped up onto the bridge. "Who is that? The only thing I can see in the boat they're towing is two cases of beer. For these times I would expect something better. I don't like this."

"Sylvie, there you go again, getting suspicious."

"That's right." She stepped back onto the deck and disappeared into the stateroom. Taking down Charlie's shotgun she jacked a shell into the chamber, slid the safety latch to the side, and stood in the shade at the door of the cabin, holding the shotgun down to her side, her finger below the trigger guard.

The errant barge swung slowly in the easy current out from Barfield Point, just below the remains of the small town of Tomato.

When the pair on the barge saw the cabin cruiser coming their way, half a mile down the river, Freddy stood and took off his shirt. "Maybe I can get the attention of this boat. Maybe someone will save us this time."

"Oh, Freddy, I hope so. My stomach is feeling worse and worse, and I think it's affecting the baby. She doesn't want to eat anymore, and she's crying all the time."

"Don't worry, honey. We'll make it. I promised you, didn't I?"

Freddy saw the jet skis coming up behind the boat. "Hey, there's some people on jet skis. If the big boat won't stop, maybe we can get some help from them."

The barge approached the cabin cruiser and the jet skis unnoticed. Freddy held his shirt still when he saw what looked like a pistol in the hand of the man on the jet ski. Then the cabin cruiser slowed to a near stop. Though the other craft was 200 yards away, he sensed some kind of confrontation. "Hey, that guy's pointing a pistol at the big boat. It looks like he's getting ready to rob those folks. Quick, get down."

He dropped to the deck and pulled Paula down to his side. "Here, lay down beside me so they can't see us. If they're robbing the big boat, they probably don't want any witnesses."

As they drifted past the jet skis and cabin cruiser, Freddy and Paula lay flat on the deck of the barge, obscured by the shadows of the trees along the west bank of the river.

Paula cried. "Freddy, just when we might get some help, this has to happen. Why? What do we do now?" The world was not fair. Anarchy reigned in the world around them making them uncertain about whether to call for help or keep quiet.

Jud closed the distance between his jet ski and the large cabin cruiser. His sister followed on the second jet ski. As he came alongside the boat and matched its speed, he pulled the chrome-plated revolver from behind his belt and pointed it at the man standing on the bridge. He yelled, "Bring your boat to a halt and put up your hands. Tell those kids to get to the back of the boat."

"What the hell are you doing?" Charlie asked, his face showing utter surprise.

"I said stop the boat or I'll shoot."

Charlie pulled back on the throttle, putting the props into idle, leaving the engine running.

"My sister and I are taking your boat, that's what we're doing. If you don't want to get hurt, just move off the bridge and get to the back of the boat." Jud drove his jet ski to the ski deck at the stern of the powerboat and reached down to tie a line from his jet ski to a stanchion on the boat.

"Jud, there's a woman who's got a shotgun," yelled Sally Mae. "She just came out of the cabin."

Jud looked up. Seeing the woman raise the firearm and point it at him, he turned the tiller hard left and gunned the jet ski to speed away, turning to fire his pistol over his shoulder at the boat. Over the sound of his jet ski, he heard a loud blast from the shotgun and saw a splash of water to his right. He leaned over and accelerated back downriver.

Looking left he saw his sister moving alongside. Another blast of the shotgun was followed by a loud crash in the boat behind her. One of the two cases of beer had been blown apart, and a hole showed in the bow of the skiff. He thought, maybe the big boat wasn't worth the price.

— 16 —
The Barfield Crevasse

Monday morning found the trio of Chris, Alex, and Tina only five miles into their journey. "Come on, you guys." Tina's hoarse whisper stirred a fragment of consciousness into the minds of her companions. "If we don't head out now they'll conscript us again for work in the cleanup crews. We said we were going to Memphis, remember?" Tina shook Alex. He stretched on the flat cardboard laid on the concrete floor, then slowly sat up. His face showed his discomfort.

Chris too sat up. He said with a groan "Oh. My back hurts. This is worse than sleeping on the cot in the Seismic Lab. More like when I had to sleep on the desk. At least the cot had some give to it. "

A small aftershock shook the garage of the abandoned house where the trio had taken refuge the night before during the rain. At least the roof kept them dry. Some spilled boxes still littered the back of the room, but no others fell from the shelving above. On the other side of the room two more couples lay cuddled together for warmth.

Keeping her voice low she said, "I found some canned food and a box of cereal under a tarp in the back of the garage. There was also a plastic jug I filled with water from the hot water heater. Thank goodness these folks strapped it to the wall; else the water would have all drained out. Alex, do you still have your pocketknife?"

Alex poked into the pocket of his jeans and nodded. "Yes, why?"

"We have to have some way to open the cans, dummy. Now come on before we wake everyone up. Keep it quiet."

Tina slipped out the back door into the early morning gray. Low clouds covered the sky, and wind blew a damp mist from the northwest. It felt cold. Chris followed.

Alex appeared at the door holding the plastic grocery bag of goods. "Here, hold the bag. I got to go."

Tina responded in a strained voice, "Well, so do I, but we have to wait until we can find someplace more private. Hold it for another five minutes. Look, there's a creek just down the road and we can use the bushes. Oops." She ducked back inside the door and returned carrying the remains of a newspaper. "Just remembered, we didn't have any toilet paper."

Another early morning thunderstorm had cooled the air and raised the humidity. Now the clouds dipped low and occasional showers pelted the refugees sitting on the barge covers. There were 100 or so survivors from Caruthersville plus another 20 they had picked up overnight. Everyone else had crowded onto the *Bella Queen.*

Buddy Joe and Barney walked the barges to double-check all the bindings. After tugging on the ties to the *Bella Queen* to assure himself they had not loosened from the humidity, Buddy Joe said, "Word must have gotten around. Looks like we picked up more refugees during the night. I assume the *Queen* is crammed to the gills, else at least some of this crew would be sleeping there."

Barney grimaced. "To the gills and more. I had to give up my quarters to a pregnant woman, who's going into labor, and her one-year-old twins. At least we found an experienced mid-wife on board."

"Barney, all those small boats tied alongside remind me of ticks on a deer in the spring. Hope they can handle it when we move. We won't be able to stop if someone gets in trouble."

Barney nodded. "Yeah, I'll be stationing some sturdy volunteers around the perimeter with life rings just in case someone capsizes. That's the best we can do. At least some of them can sleep on the barges. How many more do you think we can take?"

Buddy Joe laughed. "I figured it out this morning. We have about four feet of freeboard on the barges, and the Army Corps regulations are that we can take that down to three feet. That means we have room to add over 2,000,000 pounds, or about 6,000 refugees to these barges, just so long as they don't all rush to one side or the other. Of course, you're the one who has to worry about provisions."

Barney looked at his friend and grinned. "Most of those we've picked up didn't bring any food or water and they've already eaten all the good stuff, but we do have your supply of soy beans and corn. Hope everyone likes the taste of split pea soup and grits."

Buddy Joe took one last look. "Okay, I'm as satisfied as I'll ever be. But, Barney, do you really think it's a good idea for me to join you for breakfast?"

"You need your energy, old buddy, if you're going to get this mess down the river. Come on, there are people who want to meet you."

"I think it'll be too crowded to eat in the *Bella Queen's* galley."

Buddy Joe's fears became fact. People in the crowded dining area of the excursion boat kept him so occupied he had no chance to enjoy his meal.

Barney watched for several minutes, and then said, "You were right. Come on, let's go up to the pilothouse."

Once they were at the top of the *Queen*, Barney said, "Here, have a cup of coffee and relax. We can watch the TV news from here."

"Thanks, Barney. Say, do you have an ice cube handy?" He took the proffered morsel and cooled his coffee. "Thanks. Our ice-maker died." He stirred his coffee and took a satisfied sip.

Stepping to the pilothouse window he looked out. "I want to move downriver at 7:30, so this should be quick." Buddy Joe stared at the levee break just above Barfield. "God, what a hole."

Barney turned on the TV as the NWS logo and news anchor appeared on the screen following the last commercial of the hour. "Good morning. I'm Sherry Kent and this is the Monday morning hourly news summary from the Network Worldwide Satellite newsroom.

"The top story continues to be the great earthquake near Memphis, a 7.9 magnitude temblor in the center of the United States that happened two days ago. Authorities now estimate the earthquake and resulting floods have killed at least 50,000 people and injured 300,000 more. They do not expect to have a final count on casualties for another two weeks. There are still virtually no communications with the most devastated areas along the Mississippi from south of Memphis to north of Cairo, Illinois.

"We have new film recorded just moments ago by satellite video phone from Jerry Ribit. Jerry has been embedded with the Army First Corps since shortly after they arrived in the center of Memphis."

The screen filled with the image of a mass of people, sheltered by an occasional tarpaulin, sitting or milling on a muddy field. The Mississippi River, with low rain clouds over it, could be seen in the background.

The camera swung around in jerks to show a young man squatting on the ground next to a pickup and a camper parked in the street near the remnants of a small patch of grass. He held a microphone to a middle-aged man who sat in a folding chair with his bandaged foot propped up on the running board of the pickup.

"I am speaking with Tom Fox, a visitor to Memphis from California who arrived two days ago just as the earth started shaking. Mr. Fox, as a person from where everyone thought earthquakes were supposed to happen, what can you tell us about what you saw?"

The middle-aged man, his chin turning gray with stubble, said, "I have felt earthquakes before, but never anything like this. My wife and I came off the I-40 Bridge in our camper just as the earthquake hit. We were almost caught in the landslide that carried downtown Memphis toward the river, but we made it off the slide and drove here to Tom Lee Park. We watched several buildings along the bluffs fall from the shaking, both from the main shock and from the aftershocks."

"What are you doing now?"

"We're stuck in Memphis for the time being, but our camper has solar power and a water tank where we can purify water, so we're pretty well off, all things considered. We have a satellite phone and are helping the Mayor with communications. My wife, Judy, even talked with the President shortly after the big earthquake happened. We are also acting as a missing parent center. We've collected about 15 youngsters so far whose parents can't be located."

"Mr. Fox, how will you …" The screen flashed and went to the test pattern.

Sherry looked into the camera. "We lost the feed at that point but understand Jerry will be providing additional interviews soon."

"Continuing with the national news, after President LaPorte's declaration of Martial Law for the stricken zones in eight states, Army and Air Force units, under the direct command of General Lopez from the Joint Chiefs of Staff, have assumed full control of all governmental functions in the larger urban areas. The Army will be cooperating with the Office of Homeland Security and the Federal Emergency Management Agency.

"The Army is rushing aid into the stricken areas, but they report that all rail lines and most roads within 150 miles of the fracture zone are impassable. Airports are out of commission, though helicopter and C-130 flights are being made into the Memphis area. Power lines and pipelines have been destroyed throughout the region, cutting off all supplies of power and fuel. Officials estimate that fuel supplies have already been exhausted in the stricken areas.

"Offers of aid from countries throughout the world are pouring in and the State Department and Office of Homeland Security are coordinating the efforts of those countries.

"Financial markets are closed today and the Treasury Department says they will remain closed until further notice. Stock markets around the world have plummeted on the news of the earthquake, dropping more than 50 percent in Japan and Europe. Analysts have pointed out that as much as 10 percent of this nation's gross domestic product has been destroyed in the earthquake.

"Over the weekend the price of gold in Zurich climbed over 1,600 percent to a record high.

"At the United Nations, world leaders called for an emergency meeting of the General Assembly for Wednesday to discuss what must be done for the entire world to recover.

"The Presidential press secretary reports that the President and Governors of the affected states will meet by videophone this afternoon to map out the recovery efforts for the center of the United States. Concerns have been expressed …"

Buddy Joe reached over and muted the TV. "That's too much. The rest of the country and the world will just have to take care of itself. I have my own problems to worry about, like how to get the people on these boats to a place of safety, if such a place exists."

Barney ducked his head. "Yeah, guess I know what you mean. Sometimes it's better to focus on a tree than try to look at the forest." He looked up. "Do you want to leave a little sooner?"

Buddy Joe pushed his long body out of the chair and grabbed his coffee cup. "Yes. We'll pull anchor as soon as I get to the *Lady Bird*, say 10 minutes. Get your crews in place."

Two days ago, early in the morning, Matthew Crawley had squirmed under the chain link fence surrounding the waterworks and climbed 30 feet up the water tower ladder at Armorel. That was his favorite hangout, a place where a 12-year-old could escape the problems with his mom, and one that gave him a great view of the flat farmland to the west and Barfield and the Mississippi River to the east. When he sat up there he could let his imagination take him up and away.

Then the earthquake hit.

His perch had remained upright for less than 10 seconds. With a twist and a turn, the huge cylinder above him broke apart, showering him with the town's total fresh water supply and pulling the understructure to the ground. Luckily for Matt, the tank fell away from the side where he clung to the ladder. He rode the metal frame to a bumpy landing and got no more than a few scrapes and bruises.

His perch still rested 12 feet above the ground, too high to jump down, so he clung to the steel frame throughout the shaking, watching everything else around him fall to the ground.

Matt sat high enough to see the collapse of the levees north of Barfield near Ditch 14A. He watched the tumbling old jeep from the levee patrol tilt as the water pushed the levee to the west. The vehicle sank into the muddy water filling the ditch. He never saw another sign of the jeep or the driver. Looking south he saw other breaks in the levees.

When the shaking stopped, he remained above the ground. Around him he could see only destruction. His mother had left for work early Saturday morning and he had survived on the ladder; it just seemed safer to stay there.

People wandered about below, poking at the damage, checking for missing friends and relatives. For three hours Matt straddled his metal perch, watching the water slowly but surely envelop the town as the ground vibrated from time to time. He could see where the river had filled all the ditches to the west toward Blytheville and had started covering the new fields of young cotton and soybeans nearby.

After a time he began to think of food. Matt clambered down from his perch and ran to his house. Scavenging through the broken kitchen of his home, he packed a grocery bag with all the food it would hold, grabbed his sleeping bag and game-boy, and ran back to the water tower.

He watched as others in the town rushed to their cars and drove to the west, hoping to outrun the floods coming across the land. He waited for his mother to return.

Now, two days after the earthquake, Matt still sat above the town, watching, but he had given up waiting. The town had become an island, cut off from everything else. He realized his mother would not come back for him and worried that the flood might not retreat soon enough. His supply of food had almost disappeared and the batteries had run down in his game.

But life could still be interesting. Today he would watch a towboat in the river as it tried to avoid the hole in the levee where water still flowed out into the lands to the west.

Charlie Green stood atop the pilothouse roof of the *Amanda Blair* and held onto the antenna mast. Through his binoculars he scanned the broken levee above Barfield. He had anchored his boat 300 yards away from where water continued to pour out of the river onto the fields of eastern Arkansas.

He pulled the microphone cord of his marine radio through the side window. "Captain Simpson, I'd say the break is at least half a mile wide, maybe more. It looks like if you hold to the inside of the bend you can stay far enough away to keep from being sucked into it. I checked with the sonar. Be sure you're at least 100 yards offshore; there is too much brush and stuff next to the shoreline. I do advise that you come through at a fast pace. There's a hell of a lot of water running out that hole." He released the transmit button.

"I hear what you're saying and agree that's the best bet. We had a pretty good view of the opening from the bridge of the *Bella Queen* and I saw what looked like several small boats and some debris get sucked into the flow and out into the fields. I'd just as soon not go that way."

Buddy Joe turned and called to his first mate with the bullhorn. "Jeff, prepare to lift anchors and get underway. Coordinate with Barney and make sure his anchors are all up before we lift those on the *Lady Bird*."

Returning to the ship-to-ship radio he asked, "Charlie, did you see anything of that barge I told you about?"

"No sir. I don't see how it could have gotten past me. I'm afraid it must have been swept out this crevasse and that it's sitting somewhere in the middle of a cotton field now."

Buddy Joe shuddered. "You're probably right. But I don't think the barge will be sitting upright. Whenever the river has a crevasse like this, it cuts a really deep pool just past the levee. The water rolls over and over as it goes into that pool, that's what makes it so deep. A boat that gets sucked through the break most often yaws and capsizes. There is usually no survival. I've seen movies of it happening in past floods."

Charlie responded. "Well, Captain, just make sure your tow doesn't suffer the same fate. Good luck. We'll wait for you here below Barfield."

The flotilla of 12 barges, 29 small boats, an excursion boat, and the master towboat began moving down the river, picking up speed as it edged into the main current of the shipping channel. Jeff walked the starboard edge of the barges, checking the lines for each of the small craft that had affixed themselves to the barges during the night.

A tall fellow of perhaps 30 years stood in his boat and shouted to Jeff. "The Captain's heading directly at the levee break. Why isn't he staying closer to shore?"

Jeff ignored him and walked on.

"What's he trying to do? Doesn't he know how to steer a boat?"

Jeff called back. "Don't you worry. The Cap'n has been doing this for years and he knows what he's doing."

"He don't either. He'll get us all killed if we wash out that break."

"Don't be a Doubting Thomas. Sit down in your boat before you fall out. Just stay put and everything will be all right."

Jeff moved on down the side of the barges, talking with the people still in their small boats, calming them with explanations of what the towboat was about to do. Most men had put their families up on the barges where they sat on the sides, cheering their fathers or husbands along. To some of the children it seemed like an outing with a chance to ride on a big boat.

Sounds of activity came from behind, and Jeff turned to see the cause. He saw the Doubting Thomas untie his line from the barge and shout to those around him. "You've got to leave these barges. The Captain's driving us straight into the levee break. We don't stand a chance." Pulling his rope loose he pushed with his oar against the barge, shoving his boat farther into the current.

Jeff ran back to where the small boat had been tied. "Get back here, you damned fool. Get back." The rope still dangled in the water.

The man sat down beside his outboard engine and pulled the starting cord. The engine sputtered but did not start. He pulled it again. Finally the engine caught and the man steered away from the flotilla.

Two blasts of the horn on the towboat sounded. Jeff yelled at the man and the rest of those tied to the barges. "The Cap'n has just signaled he is steering hard to port to come around and head past the crevasse. We will be picking up speed. Watch out for the wash."

The outboard motor on the small boat coughed and died. The barges accelerated and rotated toward the small boat drifting in the current.

Matt Crawley sat on the water tower. By now the water came to within four feet of his bare feet. There was nowhere else to go.

He felt dizzy from time to time. He had eaten the candy bars he had found two days ago, and he had drunk some of the river water surrounding the fallen tower. But now he had no food and his stomach cramped.

The cage surrounding the ladder had served as a cradle to keep him from falling into the water. It had rained and thundered, but he was still alive.

When the river boats had appeared the night before he had hoped they would come to save him, but he knew that if they even tried to come to where he was they would be tumbled and would capsize getting through the break in the levee.

Matt heard the two blasts of the horn and stood so he could see what would happen next. First the boats moved toward the crevasse, and then he could see the tow turning, lining up for the bend to head downriver.

He realized that soon he would be alone again. He sat down and started to cry.

Buddy Joe looked out the starboard window of the wheel house. He saw the small boat dead in the water, coming closer and closer to the wash from the powerful propellers as the *Lady Bird Jamison* struggled to push its load down the river and past the crevasse.

"You dumb shit. You dumb shit. Why didn't you stay tied up where you were?"

The side of the towboat struck the small boat just as it came even with the stern, and it tumbled into the propeller wash. One of the crew threw a life ring out over where the small boat had disappeared, but no arm reached up to grab it.

The bow of the small craft rose once above the white froth, and then sank beneath the white water.

Buddy Joe turned his sights back downriver. Anticipating the direction he wanted the tow to go, he steered hard to starboard to stop the turning momentum and then straightened the rudder to head for the downstream channel marker.

The radio burst to life. "Now that's what I call some wild-ass boat driving. That's the way to go, Buddy Joe." Barney's voice filled the wheel house, but Buddy Joe could not clear his mind of the sight of the lone boatman being swept under the *Lady Bird* by the prop wash. But there was no other way he could have done it. No way.

— 17 —
A Monday Cruise

On the nation's political front much remained to be decided.

President LaPorte glared into the videophone and repeated what he had said only seconds before. "The area I'm talking about is shown on the maps I faxed to each of you. It covers three counties on each side of the Mississippi River from above St. Louis at Hannibal all the way to New Orleans and two counties on both sides of the Ohio River from above Louisville to its junction with the Mississippi. In addition, areas of major flooding in southern Illinois, eastern Arkansas, and southern Louisiana are included."

The President and Governors of the eight affected states had started their video-conferencing phone call 10 minutes earlier. Governor Franz expressed his opinion first. "You can't do that, Mr. President. It is unconstitutional. You are taking state possessions away without due process. It is illegal."

"Governor, let me assure you I have had my staff and the Attorney General thoroughly research this issue. Yes, it is unprecedented in peacetime. However, there are precedents from the Civil War in which the federal government took over whole states to control their recovery."

"Yes, but those were all Confederate States."

"Gentlemen." He slapped his desk. "The region we are discussing has been subjected to destruction far in excess of that caused by the Civil War. My staff tells me fatalities in the area could run as high as 100,000, though they have not all been counted as yet. There are half a million people who have been seriously injured. As least 9,000,000 people are homeless. Damage is in the hundreds of billions of dollars. Are you prepared to handle the kind of recovery that is required for that kind of catastrophe?"

"But the federal government could send the money to the states. We can handle the recovery ourselves. We would just need some help."

President LaPorte looked at the speaker and spoke in his most stern voice. "Governor, you are whining. You yourself have refused to release your own National Guard troops to help in the rescue and relief operations in the most damaged towns and cities in your own state. You've kept those troops in your capital, which I grant was severely damaged in the earthquake. I can understand your reluctance to release them, but you left me little choice except to declare Martial Law so the United States Armed Forces can go in and do the job."

He looked across the group, making eye contact with each state leader. "What I am telling you is that I plan to keep that edict in place until the recovery of the affected areas is well underway—even if it takes ten years to do it. The executive branch will work with each state to transfer control of the stricken areas over to the state as soon as possible."

He leaned back in his seat. "I have two and a half more years in this term, and I plan to run for re-election. You can start right now to find someone else if you wish, but my goal is to see to the recovery of this country. I ask that you think about that fact and consider how you can help. The area affected is a major portion of each of your states."

The Governors maintained their silence. None expressed any willingness to accept the situation. Only time would tell.

Among the ruins of Osceola, Jud and Sally Mae sat in a second-story room of the abandoned hotel, their jet skis tied to the windowsill. Water in the first story was 11 feet deep, all the way to the ceiling. Built of solid concrete, the building had withstood the intense shaking of the earthquake though it leaned two feet to the left, a result of liquefaction.

Much of the roof had fallen in, but three young ruffians still found space to sit on broken furniture in Jud and Sally Mae's room. Their boats were tied outside alongside the jet skis.

Jud spoke first. "I saw most of you guys along the river at different times this morning and I told you to find this place for this meeting. Tony, I know you from the parties down at Slam's and I invited you here to our digs when I saw you was searching through Barlett's Hardware yesterday."

"So what? Old man Barlett ain't going to be able to use any of that stuff I took." The dirty 19-year-old chug-a-lugged the remainder of the beer Jud had supplied.

"No problem. I picked up this beer from Jake's Bait Shop down the street before the water got too high, so I agree. Nobody's going to be able to use it." He looked at the next fellow. "Bud, Jake said you was working the area south of Meeman-Shelby and he had fenced some stuff for you. That right?"

"The old bastard should've kept his mouth shut." Bud pushed the mop of hair away from his face.

Jud looked at the last man, a scrawny 30-year-old with broken teeth. "Matt, you just got out of jail last week, didn't you? I thought they was going to keep you there for another five years for hijacking the sheriff's car."

Matt took a sip of his beer. "So why'd you offer us all this beer? What do you want?"

Jud rubbed the bumps on his arm and fidgeted. He couldn't sit still after taking a hit of meth earlier in the morning. He smiled at the group. "I just got to thinking. We're all wise to the ways of the world, and the world has gone to hell around us. Most everybody's been hit really hard with this earthquake, and here along the river most everyone's trying like hell to get away. They're just leaving their stuff where it lays."

"Uh-huh. So." The group chorused their agreement.

"So I think each of us has been taking advantage. We can collect whatever people have left, but I was thinking that if we worked together we could do a whole lot better."

"Yeah, how much better can it get?"

"Sally Mae and me almost got us a cabin cruiser yesterday, but the old woman pulled a shotgun on us. They've got food and money and jewelry onboard. If we all came from different sides, the old woman wouldn't stand a chance. Now I already got two jet skis and know where I can get three more. What do you think? Want to help me get that boat?"

"But what about that old woman and whoever else is on the boat?"

"Hey, there's lots of bodies floating down the river right now. Nobody would know the difference. And once we have that big boat to use as a base, there are other big boats along with some of the big homes along the slopes downriver that we can pick off. Think about it."

After a moment, Matt asked, "So if we work together, are you saying you'd be the Captain?"

"Yeah."

"What makes you such hot shit?"

"Cause it's my idea and I got the beer." He turned to his sister. "Sally Mae, give these guys another beer and get one for me and yourself. I want you tingly again."

Once the warm bottles were in their hands, Bud raised his drink on high. "I'm in." A chorus of agreement sounded. "So here's to the Captain of our pirate crew, at least so long as he don't screw up and run out of beer."

Tina, Chris, and Alex walked single file on the right side of the road heading south. Sometimes they had to push their way through the crowds that streamed north on US-Highway 51.

Tina asked a couple they met walking up the highway, "Where are you coming from?"

The woman looked at her with dull eyes and answered, "We're from Covington. Our house burned."

"Why are you heading up this way?"

"People said we should head north. They said it would be safer up here. Our car is down in the mess at the Hatchie River Bridge. The bridge fell into the river and everyone's stopped there. We just left the car and waded across the river and kept walking this way. Do you know of any place we can get food or water?"

"They were passing out some food on the other side of Dyersburg, but be careful, you might get caught in one of the cleanup work parties they're setting up before you can get there."

"Thanks." The couple moved on north.

Alex asked, "Tina, are you sure we should be heading this way? Everyone else seems to be heading north."

"Alex, we have to help the people of Memphis. And my dad and Chris's dad are there. We need to find them."

Chris asked, "Hey, do you hear a roar?"

Tina looked up startled. "What, is another earthquake about to happen?"

"No, this is more like an engine. I think there's some big truck headed the way we're going." He looked back to the north. "See, there it is coming over the hill."

Alex and Tina turned as an olive drab humvee came over the top of the last hill they had traversed, followed by a row of Army trucks. The convoy dipped down into the valley then up to where the three stood.

The convoy moved along very slowly, allowing the northward stream of humanity to part to either side of the road to let them through.

As the humvee came even with them, Tina grabbed the door jam and called. "Are you going to Memphis?"

The sergeant looked out and said, "Yes. We're a medical unit, part of the relief effort."

"Can we ride along?"

"If you want to go back to Memphis and help, sure. Everyone else seems to be leaving the area. We need able-bodied volunteers."

Ron resumed his duty as a lookout at the stern on the second deck of the *Bella Queen*. Jeff joined him from the *Lady Bird Jamison* after they cleared the Barfield crevasse. Ron asked, "How did it go, getting past the levee break at Barfield?"

Jeff leaned against the rail and chewed on a match. "Well, Cap'n Simpson is one of the best towboat skippers on this river, and he maneuvered this tow around on a dime, right around the islands next to the east shore."

"Any problems?"

Jeff spat into the water below. "There was that one fool who thought he knew better than the Cap'n how to steer a boat. But now he's history."

"I got the impression from Captain Ruggs it might be a little touchy maneuvering through this part of the river."

"It is. Course, it helped that we had the depth finder on the *Bella Queen* to help find the bottom. We were out of the shipping channel, you know. But with the river running this high, we expected to clear any sandbars anyway."

"Why is the water running so fast out the levee breaks? I didn't think the river was that much above flood stage and would have thought there would be only some minor water flow."

Jeff gave Ron a knowing smile. "That's one of our problems here on the river. For the past 70 years or so the levees have kept the river channel right where it is. Sediments have raised the base of the river, and the Army Corp and levee people just kept making the levees higher. Some of the land surrounding the river is lower than the river itself even when it's below flood stage. Only the levees and pumps have kept that land from being flooded in normal years. Now the levees are down and the pumps don't run."

"How much elevation drop do you figure we have around here?"

"Well, Captain Simpson checked with the Army Corps by radio and they said there were some lands to the west of Osceola that were over 20 feet below the level of the river. Of course, we're already above flood stage and the Kentucky Dam break has added more water to that."

"Any idea what's next?"

"Cap'n Simpson wants to anchor just outside Osceola if it's safe. The *Lady Jane Wilson* was in Osceola when the earthquake hit and he hasn't heard from Cap'n Taylor since the big aftershock. According to Charlie Green, there was like a tidal wave through the area about that time. Maybe we can find something. Anyway, there are probably lots of refugees that need help. Who knows, we may have to build a second story on this floating hotel."

"We've got much the same situation here at Osceola as at Barfield, Barney. There's a levee break north of town that's flowing pretty hard. It looks like the town and harbor must be totally flooded."

Buddy Joe had already pulled his tow close to the south shore, three miles above and across the river from the entrance to the Osceola harbor. "I plan to anchor here just off Keyes Point. Course, you can't see it now. I guess it's about 10 feet under. I see some of the structures around, but no sign of the shore, other than some lonely trees."

Barney looked out from his post at the bridge on the *Bella Queen*. He responded on the radio. "Looks like there are a few small boats waiting for us to arrive. What kind of welcome do you think we'll get?"

"My guess is that this is the first sign of help they've seen, so expect lots of refugees. I just hope the water holds out. You're keeping everyone on tight rations, aren't you?"

"Water's been restricted since yesterday, and we set up old-time marine toilets last night. No more flushing. People hang their butts out over the water, like in the old-time Navy. I hate to dump raw sewage into the river, but there isn't any choice any more. It's funny, but some folks are really objecting, afraid someone on shore will see them doing it."

"That's what comes from doing the tourist boat thing. If you had become a true-blood river man like me you wouldn't have to make such weighty decisions." Buddy Joe laughed. "Course, you're probably being paid more than me, that is if either of us ever gets paid again."

Buddy Joe leaned out the side of the wheel house and used the bullhorn to call Jeff. "Mate, get ready to drop the anchors on my call." He increased the reverse speed and watched his absolute speed as determined by the GPS drop to zero. "Okay," he called, "Drop anchors."

Thumbing the microphone he called to Barney. "Barney, have your crew drop anchors fore and aft. I'll slow the engines in two minutes. We're in about 25 feet of water. We'll rest here through the night. I'm taking a search party into Osceola Harbor."

Barney walked into the lounge as Carolyn Phelps opened the next segment of NWS News. "We now have a special report from David George on the flooding that has plagued New Orleans and Louisiana for the past two and a half days."

The TV screen showed a scene from a small boat, looking back over the outboard motor as the boat traversed what looked like a canal in Venice. Ornate buildings lined the waterway with what looked like docks reaching out into the water. Two people in festive costumes waved at the boat as it went by.

George's voice explained what the audience saw. "Sixty percent of New Orleans has become an inland sea. Water covers the first floor of these old buildings in the French Quarter, leaving our boat direct access to the second floors of the buildings, some of which are still occupied. Many people have steadfastly refused to leave their homes even though the authorities told them that this flood would continue to cover the city for weeks, possibly even months. Scenes like this can be found throughout the area."

The TV shifted to pictures of the breaks in the levees next to the city. "Damage to the levees protecting New Orleans and other low-lying cities along the Mississippi River in Louisiana occurred immediately when the earthquake struck Saturday morning. Before that, the river had been running near the top of the levees from the spring floods coming from the melting snows in the northern reaches of the United States. After the earthquake, 17 major breaks were reported along the river below Baton Rouge, and the flooding came without warning. Within 15 minutes water had covered the heart of New Orleans."

"Authorities unofficially place the death toll in New Orleans alone at over 7,000. There are 30,000 people injured and 400,000 people have been displaced from their homes.

The state of Louisiana expects the death toll from both the flood and earthquake to exceed 35,000. They say there are 2,200,000 people without homes in the state at this point. Authorities are still trying to work out evacuation routes to move these people to dry land in neighboring Texas and Mississippi."

"As the water spread out across the land and much of the Mississippi River flow was diverted into the Achafalaya Basin, the river level from Baton Rouge to New Orleans has dropped over 13 feet. Over 60 percent of the land near the river is covered by water from one to ten feet in depth. Areas near or below sea level are covered even more deeply." Aerial footage on the TV screen showed vast reaches of New Orleans and surrounding lands. It looked like a large, shallow lake with buildings scattered across its bottom.

"On the Achafalaya, the river flooding has taken out all the roads, power, and bridges. Morgan City and numerous other small towns simply disappeared under the swollen river waters. There are no comments from the Army Corps on what is planned for that area."

The camera returned to George. He stood next to the state capitol in Baton Rouge. "Already, there are cries of rage from politicians. Some are blaming the Army Corps of Engineers for simply building the levees higher and higher, so that a disaster such as this could happen. Others are crying foul as the federal government moves Army units in to take control of the flooded areas. Still others are asking for more federal aid to help the state recover. There does not appear to be any organized approach at the state level. It is as if it is all too big a problem for these politicians to consider. They are in shock, like many of the citizens of this state."

"This is David George reporting from Baton Rouge Island, Louisiana."

Captain Barney Ruggs of the *Bella Queen* expressed his surprise when he received the request. "I wish I could do more. You make me proud, but you see I'm not like the Captain of a ship on the high seas. This is only a river excursion boat, and I'm not empowered to marry people. I'm just not important enough." He smiled at the couple, wondering if the events of the earthquake had hurried them along.

Ron looked at Lynn. "Guess we'll just have to wait. Maybe we can find someone in Memphis when we get there."

Lynn smiled. "I'm ready to be married. I've found my man. But that's the way it is." She looked at Barney and explained. "I almost had to ask him myself, and when he finally spoke up I didn't want to let him get away. But I realize there's nothing we can do but wait."

"Thanks anyway," Ron said to Barney, a sad smile on his face. Lynn took Ron by the arm and marched him back down the stairs to the stern of the boat.

"Well, we may not be married, but that doesn't mean we can't act like we are." She wrapped her arms around her man and pulled his head to her lips. Their kiss lasted so long some of those watching wondered if the pair had turned to stone.

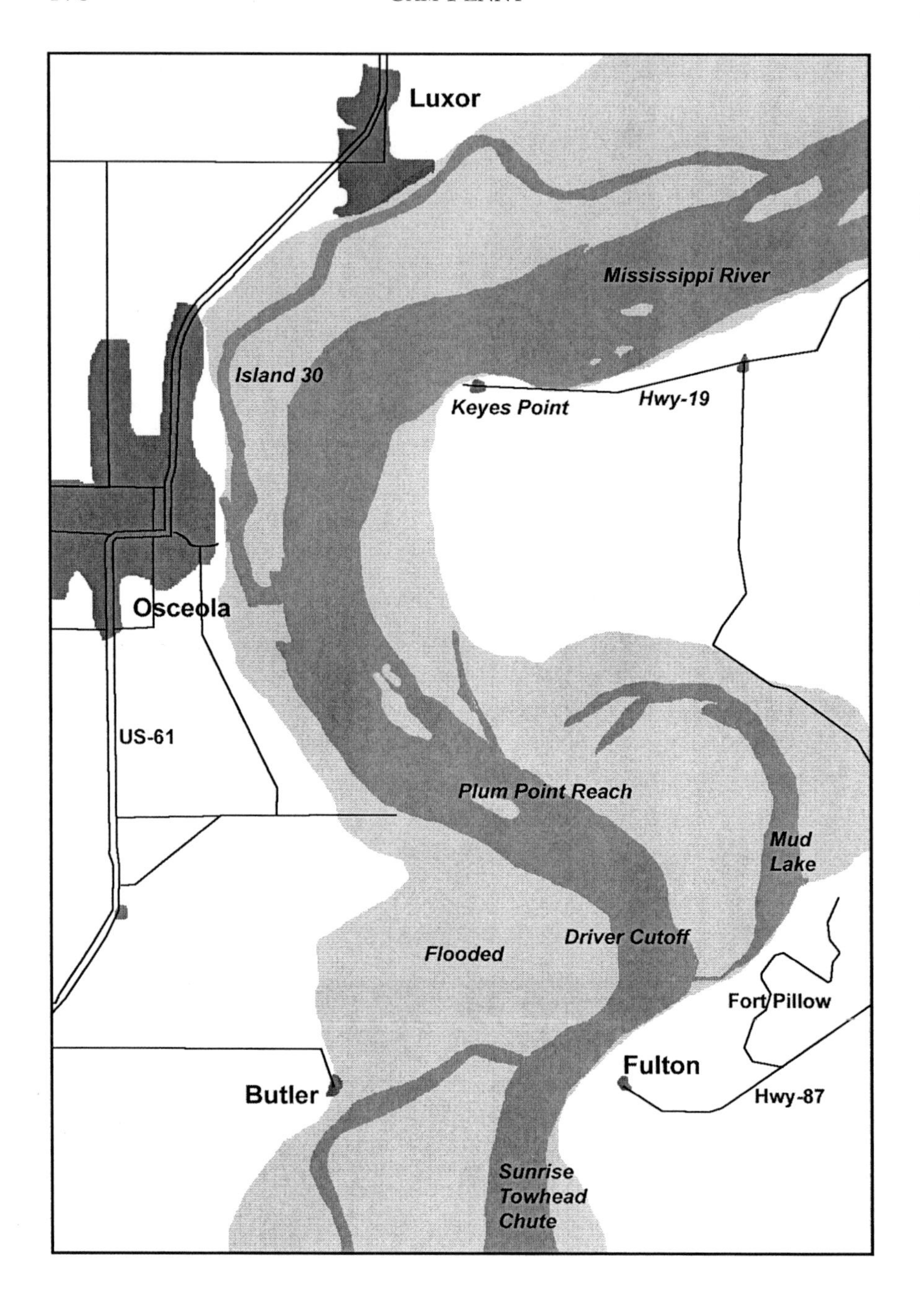

Osceola and Fort Pillow

— 18 —
Osceola Harbor

Buddy Joe leaned on the helm of Charlie Green's cabin cruiser, looking out the windows to the bow, port, and starboard. "My God, Charlie, this place looks worse than if it had been hit by a hurricane. I know we're in the middle of a flood, but nothing matches what I see here."

The *Amanda Blair* motored at quarter-speed up the channel of the Osceola estuary. Buddy Joe continued, "I saw what the hurricane surge did to New Orleans after Claudette, and it just doesn't compare to this. What happened?"

Sylvie and the McCutchen girls had willingly transferred from the *Amanda Blair* to the *Lady Bird Jamison* when Buddy Joe had whispered to them that they could use his quarters and take a shower if they wanted, but not to tell anyone. Buddy Joe and Barney joined Charlie on the cabin cruiser, leaving their respective first mates in command of the big boats. Barney agreed to bring Ralph's son, Ricky, with the admonition "You just remember now, no shenanigans."

Charlie pointed to a grain elevator lying on its side on the west side of the estuary. "I've told you about the seiche we saw from the slide at Fort Pillow. We were downstream about five miles when it came through, and I knew it would sweep up the river into Osceola, but I had no idea it would be this big. We did see waves along the shore that must have been 20 feet high."

Buddy Joe shook his head. "All the trees along here have been knocked down to the north, and those barges and boats were pushed onshore like kindling." He shook his whole body, trying to relieve the tension. "Have you seen anything of the interior of the town?"

Charlie steered to the middle of the channel before he answered. "No. I didn't want to venture too near the buildings. I wasn't sure what kind of debris might be scattered in the area, and quite honestly, I saw some looters working the area and decided I didn't want to confront any of them. I think we better be careful along that line even now."

Buddy Joe turned and called to Barney and Ricky in the stern. "You guys keep a sharp lookout. Charlie said there were some pirates operating in this area, and this boat could be a prime target." Turning back to Charlie he asked, "Have you seen any sign of a towboat through here?"

"No, sorry, I haven't."

Wrinkles grew on Buddy Joe's face. He appeared to be having a hard time coping with what he saw. "The *Lady Jane Wilson* had just delivered a set of barges to Osceola when the earthquake hit. I talked with Paul Taylor, the Captain, about the shaking and what was going on. Then I didn't hear anything more from him after the second big shaker. From what you say, that must have been the one that caused the slide."

"That's right. Are you sure he was here in the estuary?"

"Yes, he told me earlier he was tying off up ahead at the McLaughlin Steel Works dock and then taking the crew to lunch. It's hard to tell, but I believe that must be the remains of McLaughlin's main building, just above that dock." Buddy Joe pointed to a piece of steel framing poking above the waters.

Charlie continued to scan the waters of the bay. "Captain, over there I see a boat, on the far shore starboard. It's big enough to be your towboat. It's capsized and sunk low in the water." Charlie spun the wheel and gunned the motor to move across the bay. The gray hull of the heavy towboat lay fully exposed, the giant propellers and its rudders pointing skyward. The stern was raised and it rested across another boat. Charlie slowed. He could read upside-down letters on the bow that read '*Lady Jan…*', disappearing into the water.

"I'm sorry, Captain. It looks like your sister ship was destroyed by the wave. It's hard to tell how anyone could have survived."

"You're right, Charlie. The boat's gone, just like that."

Charlie steered the *Amanda Blair* around the *Lady Jane Wilson*. "Looks like a barge or something must have hit her pretty hard in the side. There's a big rip in the hull amidships."

Buddy Joe said nothing. He seemed lost in thought. Charlie could see him swallowing from time to time.

A voice called. "Hey there. Ahoy the boat." Charlie looked around, trying to find the source of the yelling. "Over here, in the towboat. Help us."

Barney yelled from the stern. "Charlie, inside the boat, in the water below the broken hull, it looks like a couple of men." Charlie steered closer to the wreck.

Buddy Joe stood up and yelled. "That's Paul." He rushed to the fantail as Charlie eased the stern around. "Paul, we see you. Are you okay?"

As the *Amanda Blair* neared the hull of the capsized towboat the man in the water dropped his arm and leaned back to lift the head of an unconscious man in a life jacket out of the water. "I'm okay, but Ben's got his foot hung up in the wreckage and I can't get him loose. He keeps passing out. Thank God, you're here."

Charlie lightly tied the bow of the *Amanda Blair* to a broken timber protruding from the hull of the *Lady Jane*. He stepped down to the deck to hold the motor yacht in place with the boat hook as Ricky and Barney donned life vests and jumped into the water.

Barney said, "Here, let us take over. You get on the boat. Ricky, try to find out how the man is caught." They worked to cut Ben loose from the tangle inside the boat while Buddy Joe and Charlie pulled Paul into the boat.

Charlie rushed back to the stateroom for a blanket. Paul shook from being in the water so long and collapsed onto the deck. "Here, Buddy Joe, make him as comfortable as you can."

Charlie returned to the side and watched as Barney held Ben's head above the muddy water and Ricky removed his life vest and dived. "How's it going?"

"Ricky's feeling down his leg for what holds it so tightly. This is the second time he has dived down."

Over a minute passed before Ricky returned to the surface, blowing out a breath and gasping for air. "His foot is caught between two pieces of timber. I can get my hand around the outside piece, but I can't get enough leverage to pull them apart."

Charlie asked, "If you tie a line to it, can we pull from here?"

Ricky sputtered,"Yes, that's probably the only way to do it."

Charlie opened one of the compartments and pulled out a length of half-inch yellow nylon rope. "Here, tie this end around the timber." He flipped the end of the rope to Ricky who took another deep breath and dove beneath the surface.

"How are you doing, Barney? Is Ben conscious?"

Barney lifted Ben's head higher and said, "My arm is getting tired holding him out of the water. He's a limp rag, but he is breathing. I think he's in pretty deep shock at this point."

Ricky popped back to the surface. "It's tied in, so you can pull on it now. Give me a minute and I'll go back down to see if it's doing any good."

After five more deep breaths Ricky yelled, "Pull," and he dove again. Charlie and Buddy Joe heaved on the line, bringing the hull of the *Amanda Blair* against the capsized tug. Even leaning back with their feet pushing against the side of the boat, they could feel no give in the line.

Ricky popped back to the surface. Once he had his breath he said, "I can feel it giving some, but not enough. You need to pull harder."

Charlie said, "The only way we can do that is to tie the line to a cleat and let the *Amanda Blair* do the pulling."

Buddy Joe said, "Let's do it. We've got to get Ben out of there, and soon."

Wrapping the line around a cleat on the far side of the deck, Buddy Joe came back to the side next to the *Lady Jane* to guide the line as Charlie stepped into the pilothouse.

"I'll steer hard to port to push the stern away from the tug. Are you ready?"

Ricky took another deep breath and dove in as Buddy Joe yelled, "Go for it."

Charlie dropped the propellers into gear and applied power. The rear of the boat moved to the side, away from the tug. Buddy Joe lifted on the yellow rope as it tightened. "More," he called. Charlie throttled up the engines and the boat moved inch by inch away from the capsized hull. Tension in the yellow rope grew, and then the tension gave way.

"He's free," yelled Barney. Ricky burst to the surface, and Charlie cut the engines.

Buddy Joe and Charlie helped Ricky back into the *Amanda Blair* and the three then pulled an unconscious Ben up the side and out of the water and lay him on the deck. Barney climbed up the rear ladder by himself.

"Oh, God, Ben's leg and foot have been crushed." Buddy Joe dropped to his knee to cut away the leg of the coveralls around the broken limb. "We've got to get him to a doctor."

Barney said, "I just hope we picked up a good one among the refugees." Turning to the boy he said, "Help Charlie cast loose and let's get back to the *Lady Bird* and the barges. And, Ricky." He reached out to shake hands. "Thanks for saving this man's life. You're a good crewman to have around."

Ron stood with the other passengers and refugees in the lounge as the six o'clock news from NWS beamed from the satellite 21,000 miles above the equator. Once again, Harold Owens occupied the anchor's chair.

"The Army has issued a notice that it has seized control of the Memphis area. There has been an increasing problem with violence and looting, and those troops that have made it to the area report that chaos and anarchy are becoming more and more prevalent.

"Meanwhile, on Capitol Hill a growing contingent of senators and representatives are calling for the impeachment of President LaPorte. The Governors, legislative bodies, and representatives from the states directly affected by the massive earthquakes that have rocked the New Madrid Fault Zone are leading a massive campaign to bring an end to what they call the 'illegal and heavy-handed' approach that the President has used in organizing rescue efforts for the stricken areas along the Mississippi River."

Paul Taylor leaned back against the pillow in the master's bunk on the towboat. "Sorry to put you out, Buddy Joe, but this sure feels nice. Better bunk than what I had on the *Lady Jane*."

Buddy Joe patted his friend and fellow Captain on the shoulder. "Well, I suppose you can use it for the time being, but don't get too used to it. I may want that bunk back some day. I can sleep in the recliner in the crew's lounge for the time being."

Buddy Joe continued, "But I want to know just what happened in Osceola. Charlie Green told me about seeing a tidal wave out on the river and how he figured it headed into the Osceola dock area. Where were you?"

Paul shuddered at the memory of what had happened two days ago. "All the barges from my tow had been tied off at the steel works when the big earthquake hit. I backed the boat off shore when the shaking began and watched the entire dock area fall to pieces from the shaking. The water in the middle was shaking around lots, but it was nothing like what was happening to the docks and buildings lining the estuary. It lasted over a minute, and then it quieted down.

"There were fires all over and people were screaming. I eased back up near the docks and started taking on people from shore. It seemed to be safest on my boat."

Buddy Joe asked, "Were there other boats in the water?"

"There were a few, but not many. Some cast off after the earthquake, but lots of the boats were damaged at the piers. There was lots of fuel on the water, and fires started pretty quickly and burned most of the boats. I had to move to the center of the estuary to avoid the fires."

Paul took a sip of water. "Then, after maybe a couple of hours we heard the sounds of another big earthquake, and the water shook again. A lot of those buildings still standing on the banks fell, and I saw a number of the people working on rescue run for their lives. That earthquake lasted maybe 50 seconds. Then it quieted down, and people went back to what they were doing."

Buddy Joe asked, "Was it the second quake that caused the tidal wave?"

"From what Charlie said, it must have been. That big slide down at Fort Pillow must have created a massive water bore. I remember seeing one like that up in Turnagain Slough in Alaska. We had no way of knowing it was coming. About five minutes after the second quake, someone yelled and pointed at the entrance of the estuary. All I saw was a wall of water rushing toward us. I swear it must have been 40 feet high and breaking over the top.

"It seemed to get higher as it got closer to us. The *Lady Jane* was turned broadside to the wave, and it just picked us up and rolled us over. I think it rolled us over at least twice. I remember I was in the pilothouse, and then everything went black. I had put on a lifejacket, and next thing I knew my head popped out of the water and I was next to the hull of the *Lady Jane*."

"What about the rest of your crew? And you said you had refugees on board, right?"

"The only person I found was Ben. Everyone else was gone. He had his lifejacket on but his leg had caught on something underwater. He kept being pulled under and he swallowed a lot of water. I swam over to him and held him up as best I could. Finally, you guys came."

Buddy Joe hung his head. "I'm sorry we lost him. He just didn't seem to have the will to come out of it after we pulled him out of the water. And his leg and foot were absolutely crushed."

"I figured that. He was pretty much gone there at the last. But at least you came by to pick us up. I don't know where I could have gone if you hadn't appeared. As far as I could tell, everything was under water and the water was rising.

"After a while the only sign of life I saw around the area were some people in small boats and jet skis. They seemed to be moving around a lot but I never caught their attention. In fact, it looked like they were probably looting the remains of some of the businesses and disabled boats."

Charlie stood on the deck of his boat as it lay moored to the *Lady Bird Jamison*. Sylvie and the girls had returned from the towboat looking refreshed and clean. Sylvie put the girls to work cleaning the rear deck with a promise to join them in another game as soon as they finished.

When Buddy Joe came out of the bunkroom, Charlie called across the gap between the boats. "Captain Simpson, I'm going to motor down toward Fort Pillow to take a look at the slide and the chute. I want to see if it has changed any since yesterday morning."

Charlie waited a moment while Buddy Joe seemed to bring his mind back to the present. "Charlie, I appreciate you taking a look at the slide area. Thanks. I'm wondering if the river is starting to cut a whole new channel through there and if there is any change in the water depths approaching the chute. Do you need one of my deckhands to go along to help?"

Charlie smiled in relief. It had become more and more difficult to run the boat by himself with Sylvie occupied with the two girls. "I would certainly appreciate that. My all-woman crew gets too busy playing Scrabble," he said.

Buddy Joe chuckled. "I have just the help for you, if you promise to bring him back. My first mate would like to take some time off this towboat. Hang on and I'll send him over."

Jeff appeared a few moments later with a large smile on his face. "Hey, thanks for thinking of me. What a relief this will be to get away from that old man." He winked at Charlie.

Buddy Joe scowled. "On second thought, you might just as well keep him. He has an attitude problem." His faced relaxed into a grin and he waved. "Have a good time."

Jeff released the lines and shoved the cabin cruiser away from the side of the towboat. "All clear."

Charlie eased the throttle forward and steered his boat away from the larger boat, taking care not to bump the growing number of small boats tied to the towboat and barges. "We're really collecting a lot of hanger-ons, aren't we?"

Jeff shook his head. "That's sure the case. I just hope the Cap'n watches the weight on the barges. You know, if you pile enough bodies on these barges that are already full of grain, you can sink them."

Charlie eased the throttle forward. They cleared the last of the small boats streaming from the Osceola harbor and headed out into the river channel.

The barge with Freddy, Paula, and the baby aboard drifted on the east side of the two-mile wide eddy of the river upstream from the earth slide at Fort Pillow. It made three loops around the eddy before underwater currents threw the barge back into the main current and it started moving closer to the rapids flowing down beside the west bank.

"Freddy, do you think those people on the jet skis will bother us if we get close to the far side?" Paula had watched as a group of five jet skiers accosted boats that had come around the bend, drifting down toward the rapids. She had seen and heard them demanding money and other valuables including water, food and fuel. Then they disabled the engines of the water craft and pushed them farther into the current of the rapids.

"I'm afraid so, Paula," Freddy replied as he looked over the edge of the barge. They lay on the upper deck of the empty barge, hidden from those on the surface of the water by the small lip at the edge of the barge. "Course, they're not going to disable this barge. It doesn't have anything to disable."

They watched as another boat was accosted and were surprised when one of the skiers grabbed a teenage girl under his arm and drove off with her toward the bank.

"They're crazy," Paula exclaimed.

"I agree, but if they find us here now, they're going to know we saw what happened. That makes us dead meat."

The pair continued to watch the actions on the water, horrified at the depravity of the pirates.

Barney looked out the window of the *Queen's* wheel house at the stream of small boats coming out of Osceola Harbor toward the *Bella Queen* and the barges. "Buddy Joe, where are all those people coming from?"

"Charlie and I talked with a couple of people in boats when we were searching the harbor and mentioned that the *Lady Bird* and the *Queen* were anchored upstream and across the river from the harbor entrance. Nobody knew we were here before then. I guess the word got around, and some people organized a ferry service from town. Now everyone left in Osceola is trying to come aboard."

Barney leaned against the back wall of the pilothouse on the *Bella Queen*. Buddy Joe occupied the Captain's chair and asked, "How many do you think the *Queen* can take?"

Barney shook his head. "I'm not sure how many we can do in an emergency, but the boat is only rated for 155 passengers and crew. There won't be a problem with where to put them, they can stand up, but there is a problem with the weight. If you figure 15 people a ton, our load capacity is another 30 tons, so we can handle maybe 450."

Buddy Joe nodded. "That's about what I thought. We've got to be careful about overloading your boat. I figure I can put maybe a 150 on each barge plus 50 on the *Lady Bird's* deck, so we are looking at a maximum of 1,850 there, or a total of about 2,300. And that's when it becomes standing room only." He looked at Barney. "The problem is I'm afraid there are more than that many people in the Osceola area who will want to get on board."

Barney pushed away from the wall. "That's just the first problem, Buddy Joe. We know we can feed them your soy beans and corn, but what are they going to drink?"

"Well, you're the expert on taking care of tourists, so I put that in your hands. I'll stick to piloting the ship."

Buddy Joe glanced out the side window to the other side of the barges. "Oh, oh. I better get down there. Someone's trying to bring her bedroom suite onboard. We've got to make sure we minimize freight and maximize life."

Jud sat on his jet ski holding onto the leafy branch of a giant maple immersed in 20 feet of floodwater on the side of the Mississippi River across from Fort Pillow. Sally Mae rested in the outboard that was filled with sundry goods taken from the boats they had accosted at the head of the rapids.

"Now, why the hell did Matt have to go back in the bushes with that girl again? We had our fun with her." The other two members of his crew sat 20 feet away on their jet skis and smirked.

Tony said, "Hey, you know that Matt. His mind is just a little perverted, and that girl stuck her tongue out at him. He's back there in the brush teaching her how to do it again." He laughed.

Sally Mae snorted. "Tony, you're as sick as Matt."

Jud scowled. "Well, I wouldn't give a damn, but Matt's the one with the other gun. Now we're back down to the pistol I got at Jake's, and it only has four shells left in it. I gave my loose shells to Matt."

Bud called from his lookout position upstream. "Hey, Jud, here comes another boat, and this one is a big cabin cruiser. Maybe it's the one you tried to get."

"Can you see its name? Is it Amanda something?"

"Yes, it's the *Amanda Blair*, and it's coming right at us. Do we go get it?"

Jud felt the temptation to attack the boat again. "Damn that Matt. He has the gun with the ammo. Why did he have to go play with that girl?"

Sally Mae spoke up. "Jud, let's go on down to the foot of the rapids and wait for the big boat to come to us. Maybe we can even catch it in the dark and creep up on it. Matt will come along sometime."

Jud said, "Yeah, you're right. We'll get it next time." He started his jet ski and pulled out into the current. Followed by the other three, he gunned his craft out into the swift water and down the river.

"Freddy, is that a big boat coming around the bend?" Paula pointed.

"Yes, it's a big cabin cruiser. And look, those pirates on the jet skis are leaving. Now's our chance. Stand up and wave your sweater." Freddy stripped off his shirt and the two stood on the deck of the barge and waved for attention.

Charlie brought the *Amanda Blair* over into the eddy as his and Jeff's attention focused on the jet skiers rushing down the chute. Jeff called a warning. "Hey, behind us, there's a barge drifting on a course to hit our boat."

As Charlie turned and steered the boat to avoid a collision, he saw its passengers. "Jeff, there are people on that barge, and it's heading straight into that chute pretty fast. From the looks of it, it won't be a fun ride."

Jeff said, "I think that's the barge we saw go by with the couple and the baby. We better get them off of that thing pronto."

Charlie Green pulled along side the barge and looked up to the deck seven feet above. He called, "Do you have a rope or something to slide down? I can hold close to the barge, but you're going to have to find your own way down."

Paula looked down in fear, clutching her young daughter to her breast. "My baby, can you catch my baby?"

Jeff looked up with apprehension. "I'll try." He held out his arms and Paula carefully pitched the naked, crying babe down to him. He caught it like he had caught footballs back in high school and carefully handed it to Sylvie.

Freddy yelled, "We'll jump into the river, then you can pick us up. Okay?"

"Yes, but hurry up and jump. We're getting awfully close to the rapids."

The two held hands and jumped together into the muddy water below. Jeff threw the lifesaver to them. The pair grabbed hold and he pulled them to the back of the boat. As they climbed aboard, Charlie gunned the *Amanda Blair* back into the safety of the eddy.

They all turned and watched the barge as it swung around in the current and slid into the chute. It rushed around in a circle before the bow caught in a tangle of tree limbs protruding from the edge of the slide and hung there. The barge tilted and rolled in the current, spilling the a few remaining rocks and mud from the bottom of the barge into the river.

Sylvie looked at the naked baby, its dried umbilical cord still hanging from its abdomen. "This baby must only be a day or two old. Did you just have it?" she asked Paula.

Paula looked ready to cry. "She came three days ago. I was all alone. We've been on the barge ever since. We haven't had anything to eat and we've only had river water to drink."

"Oh, you poor children. There're some crackers, milk and cheese in the fridge that you can have. Then go into the stateroom and get yourselves cleaned up. My robe will fit you." Sylvie beamed like she had found heaven.

— 19 —
FORT PILLOW RAPIDS

The Tuesday morning sun had risen over the hills east of the Mississippi across from Osceola. Since first light the procession of boats from the stricken town had resumed, bringing the survivors whether sick, injured, or just homeless to the excursion boat and the barges moored across the river.

Buddy Joe put the bullhorn to his mouth once again and repeated his warning to the boaters. "We are taking only the sick and wounded onboard these vessels. There is not enough room to take those who are able to fend for themselves. Stay in your boats. I repeat, stay in your boats. We will be casting off soon to continue downriver to Memphis. You can travel with us, but you cannot come aboard." He could see Jeff walking the barges trying to keep people from clambering aboard.

"Buddy Joe?" The radio sounded behind him. Barney's voice from the *Bella Queen* said, "We've got a guy with his family in a boat that he says is out of fuel and he wants to tie alongside to the *Queen* for a tow. What do you think?"

Buddy Joe grabbed the mike and replied, "Barney, I guess we have to allow them to do that, but they have to be responsible for keeping their boat clear of the wakes from the larger boats. We'll only be doing three knots, so it shouldn't be too bad. But they can't come onboard. Make sure they understand that. God, from up here these barges look like an anthill that someone stirred with a stick." He pondered the analogy with wry amusement. At least this anthill offered safety for some on the river.

He estimated the number of pleasure boats, small fishing boats, and rafts milling around the barges and the *Queen*. There seemed to be more and more coming from the shore. "Barney, we've got to get out of here before these boats sink from the weight of the added passengers. Forget what I said about 6,000. Lift your anchors. I'm getting underway in three minutes." Buddy Joe reached up and pulled the cord on the compressed air horn, sounding three blasts to warn the boaters around the towboat to stand clear. Jeff came running back to the towboat.

"Engines reverse one quarter. Prepare to lift the anchors." Buddy Joe shouted the instructions to his crew as the towboat shuddered in the waters of the mighty Mississippi.

Moments later Jeff shouted. "Anchors stowed."

Pulling on the cord, he sounded the horn for a long five seconds. "Engines reverse full." He steered the rudder to pull the towboat and its flotilla away from shore and into the current. Once the towboat reached 100 yards off shore, he sounded the horn again. "Engines reverse one quarter." As the current took control, the flotilla began its slow journey around Keyes Point and headed south down the river. Small boats and rafts surrounded it like a swarm of bees.

Ron leaned on the railing of the upper deck at the stern of the *Bella Queen*. Standing next to him stood Jeff, the lookout from the *Lady Bird Jamison*, who explained, "The Cap'n sent me up here to keep watch. Things are hectic all over, but the rest of the crew can control the passengers on the towboat. We've got to keep a sharp lookout for anything that could be a problem for the excursion boat, barges, or towboat as this contraption moves downriver.

Ron pointed to a white and black shape bobbing in the water. Jeff called on the handheld radio, "There's another dead cow in the water portside, Cap'n, off about 400 yards." Jeff explained to Ron. "No danger to us from a dead cow unless it gets trapped under the bow, then it stinks to high heaven, but it does give you some idea about how polluted these waters can become at times like these."

Ron scanned the line of sunken trees in the distance. "Think those trees are pushed over from the big wave, or are they that deep?" he asked.

"Probably both. The water level through here is much higher than it was upstream, and the current is not nearly so strong."

Ron asked, "And you say that's because of the slide?"

Jeff surveyed the river ahead for debris and reported a bobbing propane tank off the starboard bow. "Yes. I saw the slide at the Fort Pillow bluff yesterday when I came down with Charlie to check the chute. It's awesome."

Jeff instructed his helper. "We're coming past Plum Point Reach. In low water, it's a bar in the middle of the river. At this level, it's all underwater and should be of no concern. Of course, we don't know just what might have hung up in the mud down there, so the Captain's holding to the starboard channel to miss it. Keep a sharp lookout up ahead on the port side for any sign of snags."

Ron scanned the river. The current rolled back and forth over the area of the bar, indicating shallower water. The sun had cleared the hills to the east and the reflection on the water lessened as the fiery ball rose higher.

"I take it Fort Pillow is those higher hills over the top of the trees about two and a half miles south. Right?" asked Ron.

Jeff glanced up from the water to place himself. "Yes, the whole line of hills is called Chickasaw Bluff Number One. Fort Pillow is on top of the rightmost bluff. They had cannons up there during the Civil War to control traffic on the river." He resumed his scan of the nearby waters.

Ron stared at the bluffs. Somehow they looked different, darker than the rest of the hills. Maybe they were still in the shadows. But as he studied them with care, he came to the realization that he saw the scar of a huge wound on the side of the escarpment.

"Wow, that was some big slide. It's over half a mile wide."

Pressing the button on the handheld radio, Jeff spoke into the microphone. "Cap'n, if you haven't noticed you should now be able to see the big slide at Fort Pillow. Looks like it's over half a mile wide. Like I said, the cut looked deep enough to almost fill the river."

Buddy Joe came back. "I see it, Jeff. Lord, that's the biggest earthmoving job I've ever seen on this river."

A moment later the horn sounded a long blast and the flotilla started vibrating as the towboat increased the reverse thrust of its props to slow its progress down the river.

Buddy Joe's voice came over the radio. "Jeff, we're a mile and a half from the point of the bend in Driver Cutoff, but from here on we'll take it really slow. You said the chute begins just around the bend, so there could be some tricky currents and brush even through this part of the river."

Ron asked, "Jeff, how could a slide like that affect the river. The river's a lot wider than that slide is high."

"Yeah, but whenever there's a slide into the river, what it does is shove the mud across the bottom and push great big mud ripples up under the water out in the middle of the river. You never know until you get there just where the ripple is or how high it is. It restricts the water and the river will get much more turbulent around the slide. That's a huge slide, and the ripples run out well over halfway across the river, right across the regular shipping channel. "

"So what will the Captain do that's different from other times on the river?"

"Well, once he can see where the main channel is, he'll take it as slow as possible through that section and try to keep the barges all in a row so the tow goes straight through the chute. The big concern is yaw in the uneven current. You'll enjoy it."

Ron looked at the first mate and laughed in droll humor. "Sure, just what we need to keep our adrenaline up. A happy day in a theme park featuring running the rapids on the Mississippi River."

Charlie Green scouted ahead in the *Amanda Blair* and reported to Buddy Joe. "Captain Simpson, I've completed the pass across the river at the head of the rapids. It looks like the underwater dam comes three quarters of the way across the river from the bluffs. The chute is about 500 feet wide at the top. At this point there is some brush evident over near the east shore, but it looks clean in the center of the channel. There's at least 40 feet of water at the head of the channel, but the current looks like it must be up over 20 knots down in the chute."

Scratching his chin Buddy Joe thought about the information from Charlie. The 39-foot motor yacht had the size and power to handle the heavy currents in the chute. Buddy Joe wondered how he should drive his tow through the chute. To make that crucial decision, he knew he would need more vital information because, once he was committed to a course, there would be no turning back.

He keyed the microphone, "Thanks, Charlie. Now I would like you to make your run down the chute. I'm most interested in the speed of the current and the channel depths. And watch for any obstructions, on top or under the water. I'll stay off the radio so you just report every 15 seconds. We'll record your comments."

"Yes sir, we're underway. I'll keep my boat in reverse prop at about five knots and let the current pull us through. We're about 500 yards up from the start of the chute." Charlie used the GPS to measure the boat's speed relative to the land while Freddy watched the sonar for depth readings. "First reading is six knots and 40 feet."

"Seven and a half knots and 30 feet. No obstructions."

"Nine knots and 30 feet. Level water."

The cabin cruiser rocked from the turbulence created by the different velocities of the water rushing through the chute. The water flowed faster and faster. "Twelve knots and 22 feet. Two-foot standing waves but the chute remains level."

As they neared the narrowest portion of the rapids, the current speed and standing waves reflected from the bottom approached their maximum height. "I'm seeing 16 knots and 20 feet, now 19 knots and 35 feet." Five seconds later Charlie reported, "Four-foot standing waves. Captain, there's a big drop in the bottom near the end of the chute that's created some pretty big waves. Top speed I saw was 21 knots. The channel looks about 300 feet wide at this point. Still no obstructions."

After another 10 seconds Charlie reported, "Down to 12 knots and 40 feet. We are in the wash below the rapids and there's a lot of turbulence but the waves have all dropped off." He geared the props back to a forward direction and motored into calmer water on the starbord side below the earth slide. He could see houses a mile away on the east bank that must be part of the small town of Fulton.

"We made it. Apart from the standing waves near the end of the chute, there shouldn't be any problem. You might warn the small boats to come down near the shore to avoid the waves as much as possible."

"Thanks, Charlie." Buddy Joe used his bullhorn to communicate the news with all the small boats swarming around the barges. After a few minutes he saw several of the hardier souls start their own journey toward the chute. The remainder double-checked the lines that tied their boats to the barges.

Twenty minutes later Budd Joe sounded his horn and spoke to Charlie on the radio, "Guess it's my turn now. Looks as if it'll be like other times during floods when we had to run some tight chutes. Your data's great. This tow's a 170 feet wide with the excursion boat and the small craft hanging on the sides, and the different speed of the currents in the middle and on the sides will be pretty strong but should not be enough to tear us apart. You wait there on the starboard side and we'll wave as we go by.

Buddy Joe swallowed. "Guess it's up to me now to point this rig straight down the throat of this monster. Reminds me of a matador trying to put the sword into the bull's neck."

Buddy Joe sounded three more long blasts on the horn of the towboat and called for quarter speed forward. He swung the 420-foot-long line of excursion boat, grain barges, attached small boats, and towboat around to point directly at the chute and drifted toward its throat.

Many people standing on the decks of the excursion boat and on the bow of the barges considered the trip to be a mild carnival ride. They could sense the speed with which the boats moved, and a mild side-to-side movement rocked them as the boat slipped over the first of the standing waves. It seemed like the proverbial walk in the park.

The steel ropes holding the tow together suffered a terrific strain as the barges accelerated through the chute. They squealed in protest and snapped around. The edges of the barges rubbed back and forth over a foot as the forces shifted them from side to side. The ropes held.

In the wheel house of the *Lady Bird Jamison,* Buddy Joe felt a massive cramp in his back as he anticipated the force of the water trying to throw the strange craft he piloted out of alignment with the chute. If the currents turned the articulated craft sideways in the chute like they wanted, the force of the water would rip out the middle of the tow. Buddy Joe whipped the rudder back and forth, like a master juggler, maintaining the flotilla on a straight course.

All of this action took place in relative quiet. The currents of the river made little sound and the engines of the towboat idled at quarter speed with the props reversed to apply braking to the tow.

Suddenly the scream of a jet ski distracted Buddy Joe. It raced up behind the towboat and passed on the starboard side. As it rushed ahead toward the excursion boat, its speed created a rooster tail that splashed a high plume of water into the side door and onto the window of the wheel house.

"Damned tourists." Buddy Joe hit the wiper button to clean his windshield. He didn't need that kind of distraction during this difficult maneuver.

He watched several of the small boats tied alongside the barges rock back and forth as the standing waves grew. Three capsized entirely, spilling their passengers into the water. Several people threw life rings in their direction.

He watched the bow of the first barge of his tow hit the large standing waves. People standing on the front of the barge appeared surprised by the jolt and the large splash of water that rose in front of their eyes. The wave doused them with muddy water.

The line of barges pushed through the standing wave, humping up and down like an inchworm, as each barge tried to remain in contact with the surface of the river. The force was too much for two of the one-inch steel ropes holding the tow together and they snapped apart, their ends lashing out through the mass of humanity huddled atop the barges.

"Oh, God, here we go again." He fought the rudder for control until he could steer the flotilla into calmer water below the rapids. "Engines full reverse." Grabbing the bullhorn he stepped outside and yelled to the deck of the towboat. "Jeff, get the anchors ready. At my word, drop them. We've got emergencies in the water and on the second row of barges. Passengers down, caught by the broken ropes."

Buddy Joe watched the GPS speed indicator drop toward zero. After a long minute the tow moved slowly enough and he called "Engines stop." To Jeff he yelled, "Anchors away and lash them in." He swung around and ran down the outer stairs, grabbing the medical supply box as he went by.

Charlie and Freddy stood atop the pilothouse of the *Amanda Blair* and watched the collection of barges and small craft start down the chute 1,000 yards upriver. The tow appeared to be having little problem, but through the binoculars Charlie could see the different hulls move in different directions from time to time.

Freddy asked, "It looks like they are doing okay, doesn't it?"

Charlie nodded and murmured, "Yes, but it's not over yet."

"Charlie, what's that coming down beside the *Bella Queen*? It's throwing up water like a speed boat."

Charlie shifted his binoculars and found the image of the jet ski speeding alongside the excursion boat. "That looks like it could be one of the pirates that attacked us the other night."

Freddy said, "Yeah, we saw what they were doing to the small boats as they came by. Those are some freaky kind of people, if you ask me."

The jet ski continued rushing down the chute and approached the portion of the chute with the standing waves. As the waves grew larger, the driver of the jet ski had trouble steering the jet ski then suddenly lost control and careened to the side. The driver flew from the craft in a summersault as the ski plowed into the last big standing wave.

Freddy exclaimed. "My God, that's a naked girl on that jet ski." The young brown body hit the water 30 yards further along and rolled along the top before sinking into the river, 200 yards from the *Amanda Blair*.

"Charlie, I think that must be the girl that we saw being taken to the far side of the river by that guy on the jet ski. She must have escaped."

Charlie shoved the throttle of his boat forward and they moved rapidly to where the girl had sunk into the water. "Freddy, get on deck with the boat hook and start looking for her. She's somewhere here close."

Paula, Sylvie, and the two girls were on the lower deck watching the action. Sylvie already had a life ring ready to throw.

"Over there, to the right."

Charlie steered in the direction indicated and saw a small white hand lift above the muddy water and fall back. "She's just under the surface right next to the boat."

Freddy dipped with the boat hook and found resistance. Pulling up he found the hook wrapped around the girl's shoulder. "Quick, grab her arms." Together he and Sylvie lifted the unconscious girl out of the water and onto the deck.

"I've done CPR. Lay her on the deck, on her side." Freddy reached with his finger into her mouth and pulled her tongue aside. Water drained out her mouth. Putting his mouth over her lips and closing her nose with his fingers, he gave her two deep rescue breaths.

Suddenly the girl coughed, took a deep breath and coughed again, then vomited onto the deck of the boat. She went into spasms of coughing and gasping, struggling to fill her lungs with air to replace the water she had inhaled from the river.

Paula brought a blanket up from the bunk room and exclaimed, "I'm sure that's the girl those river pirates took to the far bank."

The naked girl sat up and hung her head, sobbing. Paula wrapped the blanket around her and hugged and supported her. "You'll be okay, girl. You'll be okay."

Charlie turned his craft and hurried downstream to rejoin the tow. He turned to Freddy and smiled, relief on his face. "You're a pretty good guy to have around. You did a good job."

"We're meeting here in the *Lady Bird's* galley because it is the biggest private room we have. There is no privacy anywhere else. Four of you get to sit in the booth, the rest of us get to stand."

Buddy Joe stirred his cup of coffee with a stainless steel spoon, trying to cool the liquid. He missed his ice cube. He sat the cup down on the edge of the range. "Captain Paul Taylor of the *Lady Jane Wilson* is joining us. Even though he doesn't have a boat, he knows the river and I need his advice. I've made him my second-in-command."

Others around the room nodded at Paul.

"Paul, you know Jeff, my first mate. And I think you have met Barney Ruggs of the *Bella Queen* and his first mate, Ralph Robinson. Charlie Green of the *Amanda Blair* has brought along Freddy. What's your last name, Freddy?"

"Franklin. Freddy Franklin."

"Glad to have you. I also invited two military folks who are passengers on the *Queen*, Captain Lynn Browne and Ron Cannon. Captain Browne is stationed at Scott Air Force Base and Ron's a few months out of being an Air Force Major and pilot."

Buddy Joe took a sip of his coffee. It was still too hot.

"First, we lost 27 people when the cables broke coming down the rapids and there are another 50 injured. Some of them won't make it since we have no medical facilities."

He turned to Charlie. "I want to thank my excellent scout. Charlie Green has proven to be a godsend in the *Amanda Blair*, and he is rescuing people all over the place. And I guess you've adopted Freddy as your first mate, is that right Charlie?"

Charlie grinned and nodded. "Yes, Freddy's a good hand to have around. And my wife is in seventh heaven with five girls to take care of now, so she's no help any more." The assembled group laughed.

Lynn asked, "Captain, do you know the total number of people we have on board and in the attached boats."

Barney spoke up. "I tried to do a count after we anchored this afternoon. It's not totally accurate but I came up with 2,114 passengers and crew. That includes those who were killed. It does not include the people in the boats that are traveling alongside. I estimate their number to be about 700 but that is fluid because we have people joining and leaving all the time. For the most part the boaters are depending on us for support."

"Where do we stand with respect to food and water?"

"At this point there is no food other than grain we've taken from the barges to make the gruel we're passing out. It tastes great with the special river flavoring my chef is using."

The group laughed again.

Barney continued, "We have a very limited supply of drinking water and are boiling river water as fast as we can, but at this rate we will run out of propane tonight. For anything else we use water straight from the river, hopefully from the upstream side of the boat. The heads are located on the bow of the barges."

Buddy Joe cleared his throat and took another sip of coffee. It seemed to have cooled enough to begin drinking.

Charlie spoke up. "Captain Simpson, if you are going to remain at anchor here for the night, I want to take the two young girls we rescued from the slide over to the Fulton landing. Their home is near there, and they want to check on their parents. Any problem with that?"

Buddy Joe said, "No problem, that's fine, Charlie. We will be here at the top of the Sunrise Towhead Chute until about eight tomorrow morning, so try to make it back by then if you want to travel with us."

He turned to the rest of the group.

"On the latest news. I just talked by radio with Army Colonel Brad Bingham in Memphis. He is in command of the armed forces in the Memphis. He now knows we are up here and is looking forward to our arriving in Memphis. In particular, he wants the grain I have in the barges.

"Colonel Bingham mentioned they are having an increasing problem with looting in Memphis and I told him about our river pirates. He is concerned and warned us to be on our guard."

Ron asked, "How do we stand for getting into Memphis? And is that the place where we should be going?"

Buddy Joe took a big swig of his coffee. "Damn, it's still too hot." He put the cup back down on the stove.

Turning to Ron he said, "From here we are 40 miles upstream from the Wolf River Harbor inside Mud Island where Colonel Bingham wants us to anchor. We should be able to do that without problem tomorrow. I do not believe there are any other obstructions. He says the I-40 Bridge is still standing, though it's closed to traffic because of the collapse of its ramps. The trip should take us about six hours.

"As far as where we should go, Memphis is our destination. From what I hear, there is no place any sooner than Vicksburg farther downriver where we could put ashore and find someone who can help. This was one big damned earthquake. Everything has changed, even the river."

The group sat in silence, their thoughts focused inward on what kind of world they now faced.

"What are we going to do when we run out of beer?" Sally Mae tossed the empty bottle into the river. "There's only two more bottles left."

Jud belched and tossed his bottle after his sister's. Reaching for another beer he said, "Now there's only one." He twisted the cap off the longneck. "Hell, I don't know. Maybe we can find some more back in Osceola. Course, most of it is 20 feet deep, and I don't feel much like diving for beer, especially in this water. It stinks too much."

Tony tossed another piece of two-by-four onto the fire they had built high on the riverbank. They were seven miles below where the *Lady Bird* lay at anchor with its brood of small boats just below the rapids.

Bud said, "Gimme that beer. You guys are hogging it."

Tony pulled out a shiny revolver. "See what I found this afternoon. Now you guys don't want to argue about who gets the last beer do you?" He reached over to take the last bottle. "She's right. What are we going to do now that we've run out of beer?"

Jud said, "Maybe they have some beer on that cabin cruiser I told you about. I bet it's somewhere up there around that towboat we saw. We could sneak up there tonight after the moon comes up and pick it off." He rubbed the speed bumps on his arm. What he really wanted was another hit of meth. He could feel the shaking coming on, but he didn't know where he could find any more.

Bud stood up in the flickering light from the campfire. "We might as well. None of you jerks are going to run into town for a case of beer, that's for sure."

— 20 —

Peril In The Night

Charlie steered the *Amanda Blair* along the shore near where his maps said he would find the road leading into Fulton. "This place has totally changed, Samantha. The old landing washed away years ago, but last week I know I saw the road coming down to near the water along here."

The two McCutchen sisters stood in the pilothouse of the boat with Charlie. Danielle danced and pointed at a grove of broken trees beside the bank. "Over there. That pile of wood looks like the Cook's barn, but it's sitting half in the water. That's the same fire engine red that they painted it last year. The earthquake must have moved it; it's supposed to be farther away from the river."

Charlie idled the engine of the big boat and looked at the GPS display. "Okay, now I see where we are. A big section of the bank must have dropped into the water. The GPS shows that we're a couple of hundred feet inside the old shoreline. That explains all the brush in the water through here. It's tree tops."

"Will we be able to land, Uncle Charlie?" Samantha asked as she put her hand on his arm.

"We have to be careful, but I think so. I just need to find a solid place along the shore." He called to the back of the boat. "Freddy. Sylvie. Keep watch with those boat hooks. There's a lot of brush through here, and I don't want to foul the props. Push us off anything big."

After several minutes of careful maneuvering, Charlie brought his boat in close to a fallen white oak tree stretching from the bank into the water. Stepping down to the deck to wrap a rope around one of the protruding limbs he asked, "Freddy, do you think you and the girls can make it to shore from here?"

Freddy pushed on the two-foot diameter trunk with the boat hook and found it solid. "Looks okay to me," the young man said.

Samantha squirmed with joy and grinned. "We can make it. It's just like the tree trunks we walk on all the time around here, and thanks, Uncle Charlie. Our house is only half a mile up the road, so we'll be really quick."

Freddy and the two girls scrambled over the side of the boat onto the rough gray bark. Carefully they made their way toward the roots of the tree that had been pulled up from the red dirt beside of the river.

Sylvie called. "Remember to look back every so often so you can find your way back. I don't want you to get lost, and do hurry. It'll be getting dark soon."

Danielle turned to smile and wave. "We promise. We'll be back in 30 minutes to let you know what we find."

Charlie looked down at his wife on the deck beside him. A worry crease crossed her brow. He could feel her sadness growing. "Sylvie, they'll be okay. Besides, the girls need their parents."

"I know, but I didn't want to lose them." She sighed. "But at least I have Paula and her baby to take care of. And that poor Nicole girl, she's still in shock from what those pirates did to her and from the water she swallowed. I have her in your bunk right now, but she's not really sleeping. She seems so bitter."

Charlie put his arm around his wife's shoulder and squeezed. "I love you, even when you make me play second fiddle to the kittens you bring home."

She laid her head on his chest and brushed her hair against his face as they watched the two girls leading the way toward the road to the east. He wrinkled his nose; the gray strands tickled and smelled of hair spray.

Over an hour passed before Freddy returned to the boat docked against the tree trunk. He found the going harder than he thought it would be. "Charlie, it's getting really hard to see. Could you give me a light?" Charlie shined a flashlight at the roots of the tree to show Freddy the way.

Back on deck, Freddy related his experiences. "Sorry I took so long, Charlie, but we found the girl's home. It's a total mess. Their mom and dad are camped out in the yard in a tent. They're all dirty and messed up a bit but they looked to be okay. That was sure one happy family to be back together."

Sylvie's voice showed her concern as she asked, "Why didn't they come back with you?"

"Mr. McCutchen said they had food enough to last where they were, what with their garden which didn't seem at all bothered by the earth shaking, and they were putting things back together on their farm. He said he thought they would be better off away from people, just taking care of themselves. He seems pretty well set up where he is."

Charlie said, "Well, it's his decision."

Freddy's sardonic laugh told much. "Actually, it was Mrs. McCutchen who decided. After she listened to Sam and Dan tell what we saw around Osceola, she said the world had gone to hell and the family would just stay right where they were until it all returned to normal."

Charlie's sarcastic laugh escaped before he could stop it. "Normal? What does she think that is? When does she think that will ever happen?"

Freddy waited for Charlie to calm. "Charlie, they may get back to normal before any of the rest of us do." He then asked, "So, what do we do now?"

Charlie released a sigh of tension as he looked around in the dusk. "I don't want to stay this close in. There's still enough light, so I'm going to move out about 100 yards and anchor down a quarter mile for the night. No use trying to cross the channel back to the towboat until tomorrow morning."

Freddy hesitated, and then asked, "Charlie, I've admired that awesome sound system you have on this boat. Do you mind if I play some music. It would be nice to hear some good blues, something to take us away from all this wreckage and back to how good it used to be. Besides, it might help you and me calm down."

Charlie looked at the young man and smiled. "Freddy, you are wise beyond your years. Sure, let's have some music, and I'll even fire up the gas barbeque for supper. We have some frozen hamburger patties and some buns that will go bad if we don't eat them pretty soon."

Freddy clapped. "Paula, guess what? We're going to have a party."

Jud kept his jet ski at a low idle as he led his group up the east bank of the river. Three other jet skis paralleled his course and Sally Mae pulled the empty johnboat.

"I swear. I smell cooking," said Bud.

Two hundred yards upstream from their position Jud could see the *Amanda Blair* at anchor behind some trees that had dropped into the water, its bow pointing into the easy current at the side of the river. A gala party with music and Chinese lanterns filled the back of the boat.

"Keep quiet through here. There's enough of a roar from the chute and that music they're playing should be loud enough so they won't hear the skis, but they might notice our voices."

The malevolent crew moved 50 yards closer. Jud pointed and said, "Tony, you and Bud move on up to get in close, then come in on the sides when you hear me gun my engine. Sally Mae and me will come in at the back of the boat."

The group spread for their attack in the growing darkness.

Charlie looked up as Sylvie said, "Charlie, do you hear something?" She stood by the railing of the boat turning her ears away from the music and then edged toward the stairs to the stateroom.

Charlie cocked his head to listen as he stood up from the grill and turned his head. "No. What did you hear?" The sudden roar of a marine engine filled the air, and he looked into the darkness to the south. "My God, that sounds like those jet skis."

Dropping his tongs, he raced to the pilot deck and grabbed the microphone of the radio. "Captain Simpson. Captain. Those pirates on the jet skis are attacking us. This is the *Amanda Blair*. Do you hear me? We are being attacked just off from Fulton."

Jud's ski boat bumped into the back of the *Amanda Blair*, and Jud jumped off onto the stern, holding a bright shiny automatic pistol in his hand. He wrapped the tie rope around a stanchion and shouted. "Everybody drop, get down on the deck. Now."

Charlie screamed into the mike. "Yes, we are being boarded."

Jud fired a shot through the windshield of the boat, next to Charlie's head. Charlie dropped the microphone and dove for the deck.

Two more jet skis bumped against the boat as Tony and Bud arrived. They tied on and clambered over the side.

Sylvie emerged, running up the steps from the stateroom below, holding the shotgun in her hands. She brought the weapon to her shoulder and aimed it at Jud.

"No you don't, lady." Tony grabbed the shotgun and lifted it as Sylvie fired a single shot into the air. He wrenched the gun from her hands and kicked her back down the steps.

Bud walked over and placed a foot on Charlie's neck. "You just hold still, now. No use for you to struggle."

Charlie turned his head to the side and could see Jud pointing his handgun toward Freddy and Paula and the baby. "You folks just stay quiet. Don't move and you'll be okay." Jud called back over his shoulder. "Okay, Sally Mae, bring the johnboat up and tie it in."

Charlie hoped someone had heard his cry for help.

As soon as Ron tossed his cards to the center of the table, Jeff threw three red chips onto the pile and said, "Call you, Buddy Joe. Let's see what you've got."

Buddy Joe smiled and slowly laid down his five cards. "Aces over eights. How do those look?"

Jeff laughed. "The dead man's hand. You thought that would beat me?" Disgusted he threw his cards onto the pile. "You're right."

As Buddy Joe racked the chips, the radio came to life. "Captain Simpson. Captain. Those pirates on the jet skis are attacking us. This is the *Amanda Blair*. Do you hear me? We are being attacked just off from Fulton."

The trio sat upright at the table in the galley, rigid. More sounds could be heard in the background, then the transmission ceased.

Jeff jumped to his feet. "That was Charlie Green," he said. "Did he say they were being attacked?"

Buddy Joe stood and bumped his head on the ceiling. "That's what I heard. Charlie was taking those girls to their home in Fulton. It's just across the river and down half a mile."

Jeff grabbed Ron's arm. "Ron, come on. I traded a pair of boots for one of the boats tied alongside. We can get there in five minutes."

Ron watched Jeff jerk the starter cord on the outboard. The engine sputtered into life and Jeff commanded, "Push us off from the barge."

Ron pushed the oar against the hull of the barge as Jeff turned the 14-foot outboard and gunned the engine to head into the darkness across the river. As his eyes adjusted, the light from the rising moon showed the shadow of the opposite bank.

"Do you know where you're going?" called Ron.

Jeff scanned ahead. "I know where Fulton is relative to the bluffs, so I'm going in the right direction. I figure Charlie has some lights on his boat, so we should be able to see it pretty quick."

After a couple of minutes Ron pointed and said, "There, to the right. That looks like a big boat on the water. It's about 500 yards more."

Jeff steered his craft toward the boat and throttled back. "You're right. That's it."

"So what are we going to do? Do you have a plan?"

Jeff thought a moment and then replied. "I'm not sure. Hit them with oars, I guess." He slowed the outboard even more as they approached the *Amanda Blair*. "I'll come in on the bow. Maybe we can get on board before anyone sees us." He cut the engine as the current carried them toward the anchor ropes of the bigger boat.

"Do you think this is the right thing to do?" Ron reached out and snagged the *Amanda Blair's* anchor rope to swing the outboard boat around. Jeff reached out to stop the two boats from banging into one another.

Jeff whispered to Ron. "Tie us to the anchor rope. Then we'll climb aboard."

"Whatever you say, boss. I hope you know what we're doing," Ron said, trying to sound encouraged.

The pair stood up in their boat and reached to grab the rail along the bow of the *Amanda Blair* and then pulled themselves up the side, trying to minimize the rock of the boat. Jeff whispered, "I'll go down this side deck. You take the far side."

Jeff made his way along the side deck from the bow, his rubber-soled shoes making small scuffing sounds as he crept along the narrow walkway toward the back of the boat.

He could hear voices from the rear of the boat, but it was unclear what was being said. It sounded like there were two or three intruders talking among themselves. He recognized the soft voice of Freddy speaking as if trying to keep tempers from exploding.

Moving with care, he peeked around the edge of the transom. He saw Charlie sprawled facedown on the deck and Freddy pushed back into a corner against the stern. Two thugs stood over the men. Sylvie and the girls must be in the stateroom.

The one he would come to know as Tony, the one closest to Freddy, held a chrome-plated handgun. He pointed it at Charlie's head and waved it back and forth. The other, Jud, kicked Charlie in the side. "Where do you keep your money, old man? Tell us so I don't have to kick you again." He kicked again, even harder.

Charlie grunted and moaned, grabbing his ribs. Jeff could almost feel Charlie's pain, and his anger exploded. He rose to a crouch and launched his body in a flying tackle. He hit Jud in the middle of the back with his shoulder, knocking him into Tony standing at the back of the boat. The two almost fell over the back of the boat into the water, but caught themselves and regained the safety of the deck.

Jeff bounced off of Jud into the corner with Freddy. Jumping to his feet, he waded into the tangle and grabbed the hand holding the gun, lifting it skyward. "You bastards have had it," he swore, swinging his left fist at the face that appeared over the shoulder of the man he fought. The crunch of a breaking nose sounded satisfying. Jud screamed and fell back onto the side of the boat, writhing from the pain in his face.

Bud brought his left hand up and grabbed Jeff's arm, trying to tear away the hand that gripped his wrist. Jeff cocked his left arm back to strike at Bud.

Jeff's sudden attack had surprised Ron on the left side of the cockpit. Reacting, Ron ran forward and took the long step down to the deck, his eyes fixed on the struggle. He had no warning when Bud hit him in the back of the neck with the shotgun, knocking him unconscious and dropping him face down onto the deck.

Freddy, too, had fixated on the fight and started to rise to go to Jeff's aid. He dodged back out of the way as Tony and Jeff brought the hand holding the pistol down, hitting the railing of the boat and knocking the gun loose and into the water.

As Freddy rose to join Jeff, Bud yelled from the cockpit. "You assholes stop right now." He fired the shotgun from his hip into the air toward the back of the boat, between Jeff's arm and Freddy's head. Only a couple of pellets struck Jeff's arm and none hit Freddy's head, but everyone in the rear of the boat felt the loud, hot blast of gas and they all dropped to the deck for safety. Only Jud remained where he was, leaning over the side of the boat holding his broken face.

Bud pointed the shotgun at the prisoners and would-be rescuers. "Jud, you and Tony get over here, out of the way. Next time I'll blast a hole in any one of you who makes a move."

Jeff looked up from the deck at the winner.

Paula and Nicole sat on the stateroom bench seat beside the door. Paula watched Sally Mae rifle through the drawers of the stateroom as Sylvie lay at the head of the stateroom bed holding the baby in her arms.

Paula listened as Nicole berated Sally Mae. "Why did you just stand there? You didn't care what those freaks did to me. You're as wicked as they are." The girl had told them some of her frightening and humiliating experiences after being kidnapped from her parent's boat and dragged back to shore by the pirates. Paula could comprehend only part of her bitterness.

Sally Mae ignored her and continued her search. When the commotion on the deck started, she moved up the stairs to the deck. "You three stay here and don't cause any trouble." Sally Mae looked out the stateroom door to the back of the boat in time to see a fist punch Jud's face. "Where'd that guy come from? He can't hit Jud like that."

Sally Mae stepped up the first step to the cabin door. Nicole surprised Paula when she sprang from the seat and slammed into Sally Mae's body, knocking her into the door jam and back against the wall. Sally Mae staggered when her head struck the porthole cover.

"Quick, pull her to the back of the bed," Nicole said to Paula. "You can't give these pirates any mercy. They're evil through and through." Nicole pulled the fire extinguisher off the wall, and swinging it over her head hit Sally Mae across the side of her skull. "There. That's for standing by and watching while your friends raped me."

She hit her again. "That's for laughing."

She brought the extinguisher back to hit the girl again when Paula grabbed it and stopped her. Nicole breathed deeply, and with pure hatred in her eyes stared at Sally Mae lying on the bed. "Now there's one less of you to do your dirty work."

Paula looked at the young girl in amazement. "You shouldn't have done that, Nicole. You really shouldn't have."

Outside the cabin, Tony stumbled over Jud as he hurried to the front of the rear deck. He watched Jud push himself up from the side of the boat, a bloody hand covering the lower part of his face.

Jud walked to where Charlie still lay on the deck and pulling back his right foot, kicked Charlie once again in the ribs. "Where's your money, old man? Where's your money?" he yelled.

Tony looked over the rest of his prisoners. Freddy sat in the corner of the stern deck with the intruder on the deck beside him. "What's your name? Where'd you come from? Who's the other guy?"

"My name is Jeff Collins. I'm first mate on the *Lady Bird Jamison* anchored just across the river. My friend is Ron."

Ron lay on the deck by the pilothouse, the back of his head bleeding from the blow. He began to rouse and struggled to rise to his knees. Jud walked over and shoved him with his shoe to the side. Ron fell back to the deck.

Jeff said, "Hey, leave him alone."

Tony added, "Knock it off, Jud. Leave the guy alone. Bud clobbered him good and he's not going anywhere."

Jud turned to Tony. "Shut up. I'm running this show. Sally Mae. Get up here. Bring those women with you." Jud waited, but heard no response. "Sally Mae." He walked to the stateroom and looked in. Spying Sally Mae on the bunk, he rushed down the stairs.

Jud tore around the stateroom, shoving and kicking at the three women and forcing them through the doorway and up the stairs to the deck. "Get up there on deck, God damn you. Get out of here."

Turning, he dropped to the floor on his knees and put his hand on his sister. "Sally Mae. Sally Mae. Wake up." He pushed an eyelid up, but saw only the glazed bottom of her eyeball. The skin and bone around her eye looked bruised and bloody. He pushed the side of her face and skull and they felt mushy, like an overripe tomato. "Sally Mae." His voice choked, and he let out a racking sob.

He felt her heart. "You don't have a heart beat. What did they do to you, Sally Mae?" His face reddened and the veins stood out on his neck. He clenched his teeth and screamed. "They killed you. You're dead."

Jud stood, and putting his arms under his sister's body, he picked her up and carried her up the stairs onto the deck.

Everyone stood in silence, pulling back away from Jud. Tony wondered what would happen next.

Jud laid the body of his sister in the middle of the deck and looked at him. "They killed her. Did you know that? They killed her." He stood and reached out. "Here, give me that shotgun. I'm going to show them what killing is. They can't kill my sister."

Tony stepped back and put out his left hand. "Whoa, Jud. You don't need the shotgun. You just calm down and I'll take care of it. There's plenty of time to do what you want to these people, but you're forgetting what we came here for."

Jud starred at his partner in crime. Tony knew he only heard part of what he said, but it was enough to stop him, at least for the time being.

Tony said, "Look, why don't you and Bud start searching this place. There must be some jewels and money stashed around, and maybe you could find me a cold beer in the galley. Why don't you go get a cold beer? Okay?"

Jud's body shook and he rubbed at the bumps on his arms. He looked like he really needed another hit of meth, but Tony didn't know where to find him one. "Yeah, a beer would help. Okay, I'll find a beer."

Tony breathed a sigh of relief. He pushed the shotgun's safety back in place and released pressure on the trigger. He looked at Bud. "Glad I didn't have to blow him in half," he said.

Bud shook his head. "I know what you mean but Jud's half freaked out. He's going to kill these people no matter what we do. Just let him do it himself so we're not involved."

— 21 —
Army To The Rescue

After asking Barney if he had heard the message, Buddy Joe dialed the hailing frequency for the Army Corps river patrol. "This is Captain Buddy Joe Simpson of the *Lady Bird Jamison* anchored at the top of the Sunrise Towhead Chute calling the Army Corp cutter *H. L. Abbot.* Come back, please. We have an emergency." He waited for only a moment before the radio speakers came alive.

"Captain Simpson. This is Lieutenant Anthony Moore on the *H L Abbott.* We are just below you coming around Lookout Bar at Morgan Point. What's the problem?"

"Lieutenant, we just received a radio call on our local frequency from the motor yacht *Amanda Blair* anchored off shore from the Fulton landing. Charlie Green said river pirates had attacked them, and then the transmission was interrupted. That happened just two minutes ago. Can you help?"

The immediate response pleased Buddy Joe. "Captain, we are on our way full throttle. It's about six miles to Fulton, so we should be there in twelve minutes. We're on river patrol looking for survivors and checking into reported cases of looting and piracy. Can you tell me more?"

Buddy Joe provided details beyond the report he had given to the Army Corps earlier about the landslide at Fort Pillow and the towboat's passage down the new chute. "The *Amanda Blair* is a 39-foot motor yacht with a flying bridge and large rear deck. When you get to Fulton, you should be able to see the white boat without a problem. Charlie said they were going to spend the night there rather than venture back across the river to where we're anchored."

"Captain Simpson, consider the mission accepted. We will find the boat and arrest the pirates. The latest orders are for quick, short justice for any of those using this catastrophe as an excuse for crime, whether it's in the cities or on the river."

Buddy Joe felt satisfied. "Good hunting."

Jeff watched from his corner as Jud rushed to the helm. "Quick, turn off the lights," Jud called as he searched for the switches to the running lights. "Hear that sound? There's another boat coming up the river, and it sounds like it might be pretty big. We don't want them coming over here to bother us."

Freddy whispered to Jeff, "Hey, inside this compartment behind us. Charlie has a flare gun. Do you want it?"

Bud found the switch to the fantail lights. As he turned them off and looked across the darkened river, he said, "Yeah, I see its outline. Looks like it could be a government boat."

Tony searched for the switch to the lights around the flooring of the boat. "Where's the switch, lady?" Sylvie ignored him. "You won't say?" He took the butt of the shotgun and jammed it into the light fixtures just as Jud turned off the controls for the running lights, extinguishing the last of the lights on the *Amanda Blair*. Darkness enveloped the boat.

Jeff rolled to one side and put his arm behind his body. He whispered, "Can you hand it to me?" He felt Freddy moving about in the darkness.

Jud said in a low voice, "Everybody keep quiet. We'll just let them go by. They must be on patrol looking for survivors."

The chugging of the big diesel engine in the boat grew louder and the running lights of the boat showed it would pass about 300 yards farther out. A searchlight passed back and forth across the waters and occasionally toward the bank. The tops of the sunken tree shielded the *Amanda Blair* from the lights and hid it from the search boat.

Tony asked, "What are we going to do if they see us?"

Jeff spoke from the corner where he sat with Freddy. "You guys should just give yourself up before you get into more trouble. This land around us is having enough trouble without having to worry about the likes of looters like you."

Freddy whispered again. "Here it is." Jeff reached down and grasped the flare pistol in his hand, hiding it behind his back.

Jud answered Tony. "We've got hostages, so they won't attack us. Here, give me the shotgun." Jud jacked the shotgun to make sure a shell was in the chamber and leveled it at Jeff. "And you just shut up, asshole. You're the one who's in trouble."

Jeff heard the sound of the Army Corps boat become louder and then drop in pitch and soften as it passed the *Amanda Blair* and continued upriver.

"See, what'd I tell you. There goes the boat. Now we can get down to business and find the money this old man carries on this boat." He turned to look at Charlie. "Where do you hide it, old man?"

"I told you before. We use credit cards and don't carry much cash. You took all that I had out of my pockets when you boarded us."

Jud focused his attention on Charlie. He turned the shotgun and pointed it at Sylvie, sitting on the deck beside Paula and Nicole. "Do you want me to blow her brains out? Do you? I don't believe you, old man." He walked over and kicked Sylvie in the side of the head. She fell to the side, unconscious. He snarled, "Where do you keep your money, and does she have any jewels?" and raised the shotgun to his shoulder.

Everyone fixed their eyes on Jud, responding to the cold-blooded look on his face. They all held their breath.

Charlie pleaded. "Please. I'm telling you the truth. We don't have any cash. Sylvie's jewels are in the vanity drawer next to the bed."

Jud heard a sound behind him but ignored it. Now he knew where the jewels were located and could see that Charlie was breaking under pressure. He looked up at Bud. "Bud, go check the drawer and see what you find. If he's lying, she's dead."

As Bud stepped down to the galley level to enter the stateroom, Jeff rose to his knees in the back of the boat and put his hand on the side railing. Raising the flare pistol on the rail and pointing it skyward, he cocked and fired it in one motion.

"What the hell." Jud spun around and looked to where Jeff leaned over the side. The flare climbed 100 yards and burst, opening a small parachute to hold the brilliant ball of burning phosphor on its slow journey back to the water. The light illuminated the area, clearly showing the *Amanda Blair*.

The shotgun already rested against his shoulder, and Jud fired point-blank into Jeff's back from seven feet away. He watched Jeff pitch forward against the stern from the force of the shot, and smiled as a growing pool of blackness illuminated by the eerie light of the flare covered the deck beneath the body.

"No, no. You can't do that." Jud turned back as Charlie lunged from the other side of the boat and knocked Jud against the transom. Charlie's right hand came up and pushed Jud's broken nose to the other side of his face. His left hand grabbed at the shotgun and pushed it over the side of the boat and out of Jud's grasp. It fell into the water and sank.

Jud screamed from the pain in his nose. He grabbed his face and tried to stop the renewed flow of blood from his nose, then staggered to the back of the boat. "Come on. Let's get out of here," he yelled through his bloody hands.

Aboard the Army Corps boat the Lieutenant said, "The Fulton road is just ahead. Captain Simpson said the *Amanda Blair* was somewhere near that point, so keep your eyes open."

The four men available for lookout duty lined the railing of the cutter, one manning the searchlight on the second deck. "We should be able to see their running lights from here."

The lookout nearest the stern saw the trace of the flare fly into the sky in the side of his vision and turned to look south as the flare exploded with a brilliant white light. It clearly showed they had passed the motor yacht lying at anchor 500 yards back, a short distance off shore. "Lieutenant, there she is, starboard off the stern." A second bright flash of light came from the back of the darkened boat.

The Lieutenant steered the cutter hard to starboard and gunned the engines of the Army Corps boat. A second later the sound of a shotgun blast could be heard.

"Prepare to board on the port side." The Lieutenant drove the cutter hard toward the *Amanda Blair* then reversed props to expertly bring it to a halt 20 yards away. The bright searchlight shined on the crowd standing in the back of the big boat. His loud voice came over the bullhorn. "This is the Army. We have a report of this boat being boarded by pirates. Everyone stand and raise your hands. My men have orders to fire at any sign of resistance. Prepare to be boarded."

The Lieutenant saw one of the men at the back of the *Amanda Blair* climb astride a jet ski. The man started it and sped away, raising a rooster tail as the ski accelerated out into the river. The Lieutenant yelled his order. "Stop there. Stop or we will shoot."

As the jet ski disappeared into the darkness several of the crew raised their rifles and fired, but all appeared to miss.

The Lieutenant called to his men. "Don't bother. I think some of the culprits are still on the boat." The cutter moved close enough to the *Amanda Blair* so a corporal could jump over to its deck and secure a line to a cleat.

Three more Army crew members lined the side of the cutter, their automatic rifles at the ready. The occupants of the *Amanda Blair* stood backed to the far side with their hands held high.

The older man called out. "Lieutenant, I'm Charlie Green, owner of this boat. Thank you for saving us. Those two men at the stern are pirates. The third one, Jud, escaped on the jet ski. We have two men and my wife down on the deck. Jeff, the man at the stern, was shot in the back with a shotgun after he fired the flare. The other was clubbed unconscious."

Two more men from the cutter climbed over the railing into the *Amanda Blair* and moved to check Jeff, lying still in a pool of blood. "He has no pulse. He appears to be dead." They moved to minister to Ron and Sylvie.

The Lieutenant left the helm to his executive officer as he came down the stairs. "Mr. Green, let me see your identification just to be sure." He compared the picture with the person and handed it back. "Are those in the back the only two?"

Charlie pointed back into the galley. "There's a girl on the floor of the bridge who is apparently dead as well. She was one of the pirates and struck her head during part of the struggle."

The Lieutenant instructed one of his men. "Put those two under arrest and lash their wrists with plastic ties." He turned to Charlie. "You go take care of your wife and yourself. You look pretty messed up. Corporal Kingsley, help Mr. Green."

He turned to the two men at the stern. "Corporal Williams, let's take these two men to our boat." The Army Corps had engaged the enemy and won.

Buddy Joe sat in the wheel house of the *Lady Bird* with Virgil, waiting for word from the Army Corps cutter.

Virgil asked, "Buddy Joe, what do you think that flare meant a while ago?"

He scratched his chin. "I would suppose they were signaling each other about something, but it's hard to tell from here. They must be pretty busy, because nobody has come on the radio to tell us what's happening."

Even as he spoke the radio came to life. "Captain Simpson, this is Charlie Green calling. Do you read me?"

Buddy Joe reached over to adjust the volume and picked up the microphone. "Hey, Charlie, your voice sounds good. Are you and your crew okay? Did Ron and Jeff make it there in the small boat?"

There was a pause before Charlie answered. "Captain, the Army Corps boat arrived and saved us. They have two pirates in custody and another is dead. One other escaped. Your first mate and the young pilot made it here and if it hadn't been for Jeff firing the flare, the cutter would have gone by and we would have been killed. But Captain, … I don't know exactly how to tell you this … your first mate is dead. Jud, the pirate who got away, shot him in the back with a shotgun after Jeff fired the flare."

Buddy Joe sat in stunned silence.

The radio interrupted the silence. "Captain, are you there?"

Virgil took the microphone from Buddy Joe's limp hands. "Charlie, this is Virgil. Captain Buddy Joe is here, but right now I think he's in shock. It's good to hear your voice and know you are safe. The Lord has taken care of you, but he had to use one of his own to do it."

Jud rode the jet ski at full throttle away from the lights of the cutter. He went almost half a mile before he slowed and began to consider his options.

"I know, I'll go back upriver, up that chute. They'll have a hard time chasing me there, and maybe I can find Mark. He never came down, so he must still be there on the bank, wondering what happened to that girl that stole his jet ski. We can go into Osceola together to get some more stuff, and maybe he knows where we can find some meth."

Jud turned the ski upstream and increased his speed. Soon he saw the lights of the barges and excursion boat. Knowing that the Army Corps boat would be patrolling the east side, he opted to go around to the west of the barges ahead.

Buddy Joe stood in the wheel house of the *Lady Bird Jamison*, silent and stunned by the news about Jeff. "Virgil, that boy was becoming the son I lost." Virgil reached out and grasped his shoulder. "He could have been what Jimmy would have become, but now he's gone, too."

Virgil sighed. "Buddy Joe, sometimes the Lord just don't plan things the way you want. Sometimes you just have to do what the Lord decides is best for you."

Buddy Joe shook his head. "I know. I know. But why Jeff?" He looked down the river from the anchored towboat, its surface now reflecting bits and pieces of the moonlight. "Why Jeff?"

A disturbance on the water a quarter mile below the barges attracted his attention. "Virgil, that looks like a jet ski. I see its rooster tail. It's coming up on our starboard side. Do you think that's the pirate who escaped?"

He picked up the radio microphone. "Charlie, this is Buddy Joe. Did you say one of the pirates escaped on a jet ski?"

After a moment Charlie answered. "Yes, it was their leader, the one who shot Jeff. He headed out just as the Army boat arrived. We don't have any idea where he went, except he headed directly out into the river."

"Well, we have him in sight. He's coming this way. Looks like he wants to make it back up the chute."

Buddy Joe dropped the microphone and turned. "Come on, Virgil, we're going fishing." He led the way in a run out of the wheel house and down the stairs to the wide stern deck of the towboat. On the way he explained. "Virgil, get me your fishing rod and reel. I want to catch this one on the fly."

Virgil snatched his equipment from his boat, which was tied to the side of the towboat. The quarter pound lead sinker and double-aught treble hook hung at the end of the 100 pound test nylon line wrapped about the big level-wind reel. "You still know how to cast one of these things?" he said as he handed it to Buddy Joe.

"Virgil, you taught me, and you didn't teach no dummy."

Buddy Joe could hear the sound of the jet ski coming upriver closer and closer, just 20 yards off the side of the barges. It would pass a good 35 yards from the towboat. As he saw it appear around the edge of the last barge, he swung the rod up and out, controlling the reel under his thumb to release the line at a slow enough speed to prevent a backlash in the line. First 20, 30, 40, 50, then 55 yards of the line played off the reel before the sinker hit the surface of the river. The line still floated in the air, dropping slowly as the weight sank into the water and pulled the line down to the bottom.

Buddy Joe felt a slight, satisfying tug as the jet ski rushed through the trajectory of the line. He set the brake and lifted the rod to absorb the shock he knew would come as the rush of the jet ski rider whipped the business end of the fishing line out of the water and into an ever-tightening loop around the his neck.

Virgil chuckled as the rod tip dipped. "God works in strange ways to get even, Buddy Joe. You got him."

In the moonlight they watched the rider on the jet ski stand up, then jerk off the back of the ski into the water. The dead-man's throttle immediately stopped the ski and it fell over on its side. In the meantime, Buddy Joe wound the line in, maintaining enough tension to keep control as the slow current brought his catch back to the towboat.

Soon the flailing body of the young man came into view in the lights at the side of the towboat. "Help. Help. Save me," the man called. "I don't know why you're fishing this time of night, but your fishing line jerked me off my jet ski."

Buddy Joe brought his catch next to the hull and dropped a rope over the side where the man could hold on. He asked, "Did you just come from the big motor yacht across the river?"

"Save me. Help me out."

"I asked you a question. Did you just come from across the river?"

"Yes. There were some pirates over there and they were shooting all over. I escaped in time. Help me up."

"Did you know the man you shot in the back was my son?" Buddy Joe stared at the swimmer whose eyes grew into large globes as understanding dawned. Buddy Joe looked over his shoulder. "Virgil, lend me your bait knife."

— 22 —
On To Memphis

The Wednesday morning shadows from the Chickasaw Bluffs still covered the *Lady Bird Jamison* and its brood when Buddy Joe thumbed the microphone and spoke. "Lieutenant, we're ready to hoist anchor and head south. Thanks for all your help, and good luck upriver."

The commanding officer of the *H. L. Abbott* returned the salute. "Captain Simpson, take care going downriver. I have alerted my command that you should be there later this afternoon and they are awaiting you. Your cargo of foodstuffs is important to Memphis, and your passenger list of over 2,000 souls represents a large number of people who would not be here today if it were not for your efforts. God Speed."

Buddy Joe called Barney. "Barney, has your crew hoisted anchors yet?"

"Yes sir, we're ready and willing to be hauled around by the likes of a towboat at your command, sir." He chuckled.

"Barney, your jokes are getting stale. But keep your booties on. Memphis, here we come."

Buddy Joe signaled reverse-one-quarter to the engine room, leaned out the window of the wheel house, and yelled through the bullhorn. "Paul, as soon as the anchor lines loosen, bring the anchors up and we'll move out." It was good to have his friend as his new first mate. He just needed to find a new boat for him.

At the signal from Paul that the anchors were stowed, Buddy Joe sounded three long bursts on the horn and the flotilla began moving out into the main channel and down the river. At the latest count there were 43 small boats still tied around the edges of the barges along with the *Bella Queen*. The barges were almost invisible under the mass of humanity that occupied their surface. A long line had formed at the bow waiting their turn at the head, a term most of the passengers now well understood. Most courteously stood with their backs to the boards hung out over the front of the craft.

Buddy Joe smiled. What a day for sailing down the Mississippi.

Lynn stood inside the door to the small stateroom. Tim Warren finished the inspection of Ron's skull, neck, and back and covered him with the sheet. She asked, "How is he? Is anything broken?"

Tim turned to look through his red eyes. "He appears to be in good health, all things considered. He has a concussion, which is not good, but there does not appear to be any other damage other than bruises and contusions. I think he will regain consciousness soon but have a terrible headache for a couple of days, and then he'll be as good as new."

Lynn's shoulders slumped, and Alta Warren caught her. "Oh, thank God. I've been so worried since they brought him back from the *Amanda Blair*.

Tim came over to give Lynn a hug. "I know how you feel, but he will be okay. Trust me."

Lynn looked into his eyes. "Tim, I didn't realize you were a doctor, and when I had Ron brought here they said you were busy taking care of everyone. You've been on this boat for four days now. Have you gotten any rest?"

Tim gave a dry laugh. "Oh, I've gotten a nap now and then, mostly when I was doing an examination. This has been pretty easy compared to some rough times I've been through. This is not as bad as war, but at least in battle we generally had medications flown in. Here we've run out of all medications and are depending on the original holistic style. Clamp your teeth down on a stick for anesthesia, boil the water for antiseptic. God knows how bad it must be on land where so many more people are suffering."

Lynn returned Tim's hug. "Thank you for all you have done. We will survive."

Tim smiled. "Yes, most happy and optimistic people do survive, no matter how bad the circumstances. The quality of life is determined by how you feel about those around you, not how you feel about yourself."

Sylvie opened her eyes and looked around. She lay in the familiar bunk of the *Amanda Blair*, but something was different. She could tell from the slight rocking motion that the boat must be moving.

Turning to her left she saw Paula. "Hello. What are you doing here?" she asked.

Paula's eyes welled with tears, and she leaned over to give Sylvie a hug. "Oh, are you okay?"

"What's the matter? Why are you crying?"

Nicole stepped forward and clasped Sylvie's hand and squeezed it tight. "We were so worried."

"Why were you worried? What was wrong?"

Paula sat up. "Don't you remember, how you were kicked by that pirate, and you hit your head and we thought you were going to die?"

Sylvie thought, and then said, "No, I don't remember anything like that. I feel fine, just a little sore from something. I'm okay. But I remember a dream I had while I slept, and how I made a decision."

Paula held the hand of the old woman. She looked through clouded eyes at the woman who so resembled her own mother, listening but not really paying attention.

"Paula, I lost my son to cancer, and then my granddaughters to a broken family. I've decided I want to have another family. Will you be my daughter? Please." She looked at the girl.

Paula was totally taken by surprise. "What did you say? You want me to be your daughter? But I have a black baby, doesn't that make a difference."

Sylvie laughed. "Oh, yes it does. I forgot, I have to take Freddy as my son-in-law as well. Is that okay? Do you think that will be okay with him?"

Paula, overcome with joy, looked at the woman lying in the bed. It had taken an earthquake to make it happen, but she had found a family; she had a new mother. She laid her head across the woman's breasts and broke into uncontrolled sobs.

From his vantage point in the pilothouse of the *Bella Queen*, Barney reported what he saw to Buddy Joe. The *Queen* obscured part of the view for those in the towboat wheel house.

"Buddy Joe, as we go past the Loosahatchie Bar I will be looking really hard at the I-40 Bridge. After what happened to me up at I-255 I would just as soon not run aground again."

Buddy Joe responded. "I agree, Barney. Keep sending your report. You're my forward eyes."

"Buddy Joe, from here I can see some obvious damage to the tall buildings in Memphis. But the most obvious thing is what is missing. The Pyramid is gone, literally gone. Even with all the houses and condos on this end of Mud Island flattened, I can't see any sign of the Pyramid. Damn."

The flotilla moved slowly down the channel, passing the flooded entrance to the Loosahatchie River Canal.

"Wait, Buddy Joe, there it is, close to the ground. My God, the whole thing collapsed. It looks like the four support columns just came unattached in the middle and fell in, like a teepee falling down."

Buddy Joe responded. "Now I see it. It has a striking resemblance to Mount St. Helens after it exploded."

Barney felt nauseous as they moved on where he could look at the damage on the other side of the Pyramid.

As they neared the interstate bridge, Buddy Joe said, "The connecting roadways from the bridge to the bluffs look like they're down, and every building that's still standing and facing the river shows signs of damage."

Barney could now see the Museum on Mud Island. "Buddy Joe, it looks like the monorail must have fallen during the shaking and the buildings on Mud Island are mostly flat. But the island itself seems to be all there. Some folks were thinking it would sink into the river in an earthquake."

Buddy Joe came back. "I see what you mean. But Mud Island seems to have moved. Am I right?"

Barney commented. "Well, I can see over to the west that the I-40 causeway has fallen. It looks like the main section of the bridge pulled away from the east shore and shifted to the west, dropping the eastside connectors. The shipping channel must have moved west with it." Barney held the binoculars to his eyes and scanned the bridge. "But I don't see anything that indicates any part of the road has fallen. I remember them doing retrofits on the I-40 bridge so it could rock around during an earthquake, so that must have worked. Too bad they didn't do something about the connectors."

"Okay, I'm keeping the boat at three knots headway and the current here is about four knots. Keep the lookout on your boat ready and active and let me know if any trouble is coming up. At seven knots I won't have much time to correct." Buddy Joe gripped the rudder control even tighter as he took responsibility for steering the ship through the tight quarters. He sounded the horn in two long blasts.

The refugees on the barges and the passengers lining the rails of the *Bella Queen* all stared upward as the huge bridge passed overhead. They yelled with joy, almost like tourists seeing a new sight for the first time. Everyone on board stared at the smoke coming from President's Island and "oohed and ahed" at the broken vista of the Memphis skyline. Soon they would return to dry land, not realizing how much better off they would be if they could remain on the boats.

Buddy Joe watched as the stern of the towboat cleared the tip of Mud Island. He radioed to Barney, "I'm all clear, Barney. It's about time to start moving you over."

Buddy Joe, Barney, and the Army had agreed that they would bring the *Bella Queen* to shore at the top of Tom Lee Park, just across from the tip of Mud Island.

"Engines full reverse," Buddy Joe called as he steered the rudder hard starboard. Slowly the barges came to a halt and the entire flotilla began to move backward and toward the shoreline below the Memphis bluffs.

Buddy Joe hoped that the earth's shaking had not displaced the old tie rings embedded in the cobblestones along the banks of the river. "Barney, get busy with your crew and be sure they tie your boat properly. I will hold at this point until you get a tie and get your gangplank down."

The bow of the *Bella Queen* pointed to the north, its starboard side closest to the shore. A horde of people milled about the area where Barney wanted to set his gangplanks. His concern began to grow.

Barney called from the pilot deck with a bullhorn, "Crew man Todd, the starboard gangplank won't quite reach shore yet. Loosen the front port tie line to the barges to give us some slack." He watched as three members of the crew laboriously unwrapped one of the ropes that had held them in place next to the barges on the way down the river. They had plenty of room. All the passengers stood on the starboard side, looking at the shore that lay so close.

Finally the line loosened, and the crew began to pay it out so the bow of the boat could swing closer to shore while its stern still remained tied to the barges.

Barney walked to the starboard side and looked out at the scene. "There must be 4,000 people out there next to the boat. It must have been hell here for the past four days," he commented to his first mate. "All the people on shore look the same. They're grimy and they all have dirty clothes, and they're beginning to act like a mob."

The large boat began to drift closer to the shore. "Crew man Phillips, don't drop the gangplank until I give the order. Understand?"

Barney was surprised to see the impatience of the mob on the bank. They appeared desperate to come aboard the boat. Already several men were wading out from the shore. They had to swim the last six feet to touch the boat. He called urgently, "Send some men to the stern to repel those attempting to board the boat." Calling to the port crew he said, "Retighten the tie line. I don't want the boat to drift too close to shore."

Barney reached inside the pilothouse for the microphone for the ship's intercom. "Ladies and gentlemen, may I have your attention. You folks on the shore, I would like your attention as well. Please, listen to me." The clamoring of the mob lessened. "Please, listen to me. This is Captain Barney Ruggs of the *Bella Queen* speaking. Please, listen to me." The people quieted so he could be heard.

"The *Bella Queen* wants to drop its gangplank so the refugees and passengers onboard can disembark. But at this time we cannot allow people on shore to come aboard. This boat is already packed to the scuppers with passengers and refugees. Please do not attempt to board the boat. My crew has orders to repel anyone who tries to board the boat, either by the gangplank or through the water. Do you understand?"

A chorus of boos came from the crowd. The people on shore grew more and more unruly. They started to push and shove each other. One picked up a cobblestone from the bank and heaved it at the boat.

"I don't want people to be hurt by the gangplank when it drops. I will not lower the gangplank until you people move away from where it will come down. Please move back." He waited for a response. The crowd pushed closer and closer to the bank. Some of those next to the river were shoved and fell into the water. The clamor rose.

"Barney. Barney. This is Buddy Joe." Barney looked back into the pilothouse where the ship-to-ship radio called. He grabbed the microphone and keyed it, "Yeah, Buddy Joe. I hear you."

"Barney, don't let those people near your boat. They're a mob. A troop of soldiers is headed your way double time and they should help control the crowd. Wait until they arrive before you drop that gangplank."

Barney looked to the north and could see the Army contingent running down the shoreline. "I agree on that, Buddy Joe. I see the Army now. We're trying to keep the people back right now. We've had a couple try to swim out, but I think we can handle it."

Returning to the intercom he spoke once more to the crowd. "Please folks, move back. You've already shoved some people into the river. Move back to make room for everyone. Soldiers are coming shortly to control access to the boat. Please make room for them."

Another chorus of boos spread across the crowd, but some of those in the back started to move east toward the bluff, easing the pressure on those along the shore.

Barney and the other occupants of the boat watched in wonder as Army personnel edged people away from the shoreline. They met some resistance but most people moved away. Slowly the troopers worked their way along the shore fronting the excursion boat.

A clod of mud and grass flew through the air hitting a private on the side of the face. A chorus of protests came from those who saw the act, and they turned to find the culprit. There were cheers from the vicinity of a short scuffle in the throng. A corporal pushed a ragged young man from the crowd, throwing him bodily onto the sidewalk. Barney felt gratification to see that many strongly opposed the mob violence that ran so close to the surface.

"Captain, you can drop the gangplank now," The Lieutenant called to Barney.

"Yes, sir. Todd, slacken the tie line again. Phillips, drop the gangplank slowly and be prepared to go to shore to get all ropes secured to land."

He turned and took down the handheld microphone. "Buddy Joe, looks like we're docked."

"Fine, drop your anchors. We'll hold at this point until all the refugees on the barges have departed. They have to cross your boat to reach land."

"We'll hurry. I know you want to dock your barges."

Barney replaced the microphones and hurried down the stairwell to the first deck. He planned to take direct control of the disembarkation of his passengers. At least it would be easy to identify the passengers. They wore clean clothes and were not covered in grime.

"Are you the Captain of this vessel?" The Army Colonel in camouflage dungarees and dirty boots stood on top of the gangplank before the six-inch step down to the deck, his piercing eyes level with Barney's.

"Yes, sir. I'm Captain Barney Ruggs of the *Bella Queen*." He offered his hand.

Shaking Barney's hand perfunctorily, the Colonel looked around then stepped down onto the deck. "I'm Colonel Howard Simon. Have all your passengers departed the boat?"

"No, about half are staying on the boat. Some are injured and others have no other place to go. I also have about 60 injured refugees I haven't moved as yet. Why do you ask?" Barney began to develop a sense of dread. He followed the Colonel as the officer quickly strode down the passageway toward the stern, then up the stairs to the second deck.

The Colonel stopped, turned and stared at Barney's Adam's apple. "You're going to have to ask all but the injured to leave. We need this ship for a hospital." He turned and continued his inspection.

"But, the people left on the boat are from other places. They don't have any place to go here in Memphis. I can't ask them to leave." Barney followed the short man around the stern of the boat and up to the third deck.

The Colonel stopped and stood facing away from Barney. Then doing a crisp about-face, he looked up into Barney's eyes. "Captain Ruggs,

I am confiscating this boat for use as a hospital. We have at least 6,800 seriously wounded people lying in filth throughout the city. The surgeons in this town need this vessel to perform some of their more difficult work. I am sorry for the inconvenience to you and your passengers, but please inform everyone that they have two hours to collect their belongings and depart the ship."

Barney stood dumfounded. "Colonel, I am responsible for this ship. You can't do this."

The officer took a deep breath. "Look, Captain Ruggs. I really don't want to do it, but please understand. We have a disaster on our hands in Memphis. All 14 hospitals are heavily damaged, four are completely out of commission. The people here desperately need this ship. It's the only sanitary place that can be found. I know it's difficult for you and your passengers, but it is necessary. Can't you understand that?"

Barney suddenly realized the Colonel was a civilized and reasonable man, and that what he asked was the best of a worst situation. He felt like a fool for arguing. "Yes sir. Put it that way and I understand. I'm sorry for arguing with you, Colonel Simon."

"Thank you, Captain Ruggs. You have two hours." He again shook hands with Barney, turned and went down the double flight of stairs to the gangplank.

Barney stood and looked at the city. Smoke drifted across the ruins from the south, and people covered the small park next to the river. He and his crew would have to find some place for those who would be displaced from the boat. He knew he was not going to like this job.

The last of the people from the excursion boat streamed off the gangplank. Some held suitcases in hand, many had nothing. They all looked bewildered.

"But why can't we just stay on the boat. We left our car in Caruthersville, and we live in Dyersburg. How're we going to get back home?" The elderly lady stopped at the foot of the gangplank and pleaded with Barney.

Barney once again shook his head with resignation, "I'm sorry, ma'am. I know how you feel, but the Army is adamant that they need this boat for a hospital. If you will just wait with the rest of the passengers over by the old food stand, I will be over soon and we will try to get organized on how to help you. But right now, the Colonel says to clear the boat."

Muttering to herself, the lady moved off followed by her mute husband who simply stared around at the destruction into which they debarked.

"Thanks, Captain. Thank you so much for coming to save us." Barney looked back as another older couple with two children came down the gangplank with nothing. "We lost our daughter and son-in-law at Caruthersville, and we would have lost a lot more if you hadn't helped. God bless you." The woman reached around Barney's neck and hugged him, sobbing into his shoulder. He blushed.

"You're welcome, Ma'am. I guess we should all be glad that we're alive." Barney shook hands with the gentleman and the two children. They too murmured their thanks.

A few stragglers remained, but the crew made the rounds to hurry them along. Barney had cleared the boat as directed by the Colonel. He felt sad, but at the same time he understood the need for a place to care for the most critically wounded.

Barney walked toward the gathering of former passengers who were standing in the street above the landing. As he stepped off the curb, yet another strong aftershock pulsed through the ground. The earth shook several times and then quivered. Then it shook again.

Many of the passengers had never really felt the earth shake until now. They screamed and grabbed each other, wondering where they should run, what they should do. The ground rocked for another 10 seconds and then quieted. Afterwards, fewer walls stood on the bluffs above them.

Barney held onto the light standard. His knees shook.

"Hey, Mister Captain, that ain't nothing'. We had a really big one yesterday. It was so strong that you couldn't stand up. It was really something."

Barney looked down at an eight-year-old boy standing in the muddy grass, staring at him. He seemed to be waiting for Barney to show some sign of courage.

"Don't be scared. You get used to it, at least me and my sister are used to it. Some of the grown-ups, they still seem to be scared, but it ain't too bad after a while. Really. And the TV man says the shaking' will keep going' on and on and on. It'll never stop. That's what he said."

At the direction of the Army squad, Buddy Joe moved his tow of barges 100 yards downriver and tied them off next to the shore. Already a steady stream of refugees from the stricken city had lined up to receive their portion of the grain in the barges. For some it would be made into the first substantial meal they had had in four days.

Buddy Joe went back into the wheel house and picked up the microphone. "This is Captain Buddy Joe Simpson of the *Lady Bird Jamison* calling Army Corps control. You said to call when the barges were tied off."

He waited for a minute before the radio came back on. "Captain Simpson, this is Colonel Harrison. Could you anchor your towboat at the foot of Mud Island and join me with your first mate? My command post is up near the Memphis part of the river model on Mud Island."

Buddy Joe smiled. "Be there real soon, Colonel." Releasing the microphone button he turned to Paul. "Wonder what he wants now. But what the heck. Let's move this tug over to the island."

Within 30 minutes the towboat was moored beside Mud Island and the pair had walked up the grass to the model of the Mississippi River. Paul commented, "Hey, Buddy Joe. Looks like the Mississippi River has run dry. There's no water flowing down it."

"That's what happens, Paul. With no electricity, there's no way to run the model. Only natural water runs now. Come along." He led the way to the center of the display area.

Colonel Harrison sat at one of the wrought iron tables lining the model, sipping from a 12-ounce Coors can. When the pair walked up and introduced themselves, he offered each of them a can. "The Army Corps said they would provide fresh water in event of a disaster. In this case it comes in a Coors can. I don't know how long the supply will hold out, but at least it tastes light and is less filling." He laughed.

Buddy Joe and Paul took a seat and drank deeply from the aluminum can of water. "Yes, sir. What can we do for you?" asked Buddy Joe.

"Captain Simpson and Captain Taylor. I'm responsible for the logistics part of the recovery effort in Memphis, and it's a matter of planning as we go. No one had really prepared a plan for what to do in this kind of situation. So I am improvising."

"Sounds commendable." Buddy Joe lifted his can in salute.

"I need a navy, and a commander for that navy—someone who understands how to move things up and down the river, someone who understands barges and towboats. As you already know your parent company suffered massive damage in the earthquake up on the Hatchie River and your headquarters got caught in the toxic gas leak from Millington. There is nothing left there on which to call."

Buddy Joe hung his head, shaken by the memory of receiving the news that so many of his friends had been lost. He looked up at Paul and clapped him on the shoulder. They had had so little time for grief.

"Captain Simpson, I would like you to take command of my *ad hoc* navy in Memphis, operating under my command. You will have to find boats, Captains, and crews. My responsibility will be to find cargo for you to move and to get as much information from the Army Corps about the status of the river as I can."

Paul smiled. "Way to go. You picked a real winner, Colonel." He rubbed his hand across Buddy Joe's back.

Buddy Joe started to speak then stopped and stuttered. "Why me? There are plenty of other towboat Captains around."

Colonel Harrison straightened his back. "Captain Simpson, sometimes a person stands out as one who finds a way to make things happen. Sometimes a person makes his own luck, and the luck of those around him. I believe you are such a person. Will you take the job?"

Buddy Joe looked out across the river, swollen now by the floods and spread by the broken levees. He had lost two sons and a host of friends, and now he would have time to grieve, but here was an opportunity to go and do something, to be part of saving the land. It was something both Jimmy and Jeff would have wanted him to do.

"Thank you, sir." He held out his hand, and then dropped it. "Oh, there is one thing. I need to give you something. Can you wait a few minutes?"

Colonel Harrison had been about to accept the handshake and was surprised when the offer was withdrawn, but the request was even more surprising. "Sure. What is it?"

Buddy Joe stood. "I'll be back in ten minutes. Stay right here."

The Colonel and Paul shook their heads, wondering at the workings within Buddy Joe's mind, but they were unable to fathom the reason for his strange actions. They talked of various problems facing the people of the river and how the country could recover from such a disaster in its heartland.

Paul looked to the south. "Here he comes, and he has someone with him." Suddenly, Buddy Joe's request made sense and he stood. Colonel Harrison slowly stood as well as Buddy Joe walked up, leading a scruffy young man with his hands tied behind his back.

"Colonel, my mind has been focused on exacting retribution for the atrocious crime committed by this scoundrel. He killed my first mate, a man I had come to consider my second son. Your offer made me realize that my individual hate and desire for revenge accomplishes nothing I no longer need to take this lowlife down the river in a gunnysack and watch him drown.

"Here, you take this pirate that killed my first mate yesterday evening. I have witnesses to substantiate charges of cold-blooded murder against this man. I give him to you, and if you will take responsibility for his prosecution and punishment, I will take the task of running your navy in Memphis."

Colonel Harrison looked at the prisoner. Calling over an orderly, he explained what was to be done. Turning to Buddy Joe he held out his hand, smiled, and said, "Noted."

Epilogue

Barney Ruggs and others gathered around the back of a camper parked on Riverside Street. They watched with rapt attention the small TV sitting in the camper's doorway.

"But why are we having all these aftershocks? Don't they usually stop after a time?" Harold Dobbs asked hard questions of Dr. Peter Donald, the seismologist from Georgia Tech.

"You must understand that once the earth's crust has been broken in a giant earthquake, the stresses within the crust are left in an unbalanced condition within a very weak zone. The crust tries to readjust, and it takes time. The nature of the New Madrid Seismic Zone is to take a very long time to return to stability. This was evident from the reports of 1811 and 1812 when the aftershocks continued for much of the next decade."

"How active has the seismic zone been after this recent event?"

"Counting all the temblors we have recorded above magnitude 2.0 the rate of aftershocks dropped to an initial rate of about 4,000 per hour right after the first temblor and from there to about 2,000 per hour four days later.

"And we are still seeing high-energy events mixed in with all the small temblors. For instance, yesterday there was a magnitude 7.1 event near Dell, Arkansas, and three temblors in the magnitude 6.0 range.

"In addition, to the north on the Cottage Grove Fault we experienced a magnitude 7.5 temblor. We believe that was a separate earthquake, though undoubtedly the transfer of forces from the New Madrid Zone could have been the forcing issue."

Dr. Donald continued, "Some of the damage reports from the central United States are most disturbing. The Cottage Grove event was especially harsh in the Carbondale and Marion areas where there had already been massive failures of some of the old mines. One sinkhole doubled in size and at least 30 rescuers who were searching in the rubble for survivors were lost. In addition, St. Louis again suffered major damage from the earthquake. Some of the previously damaged buildings have fallen and one of the older bridges is now down. Operations at Scott Air Force Base were suspended for 13 hours to allow crews to repair buckled runway sections. It is now back into operation."

Harold asked, "And how are we doing today?"

"We have recorded four additional temblors in the 6.0 to 6.9 range so far this morning, two along the trace of the original rupture and two along the thrust fault section of the second 7.4 event. The strongest was a 6.9 magnitude event centered near New Simon, the epicenter of the magnitude 7.9 temblor."

"And you say this will continue. Do you have any idea for how long?"

"We cannot say for sure, but if it continues on the same trend, the rate should drop to 500 temblors per hour in about a week."

"Will there be another great earthquake? As I remember, 200 years ago another even stronger earthquake happened five weeks after the first, then the final and strongest earthquake happened two weeks after that. Are we due for a repeat of that scenario?"

"I cannot say. The seismology world simply does not know. It could be that another great earthquake will occur, or this could be the last for the time being. The best hope at this point is to pray."

"Thank you, Dr. Donald." Turning to the camera, Harold brought the interview to a close. "We have been talking with Dr. Peter Donald of Georgia Tech Seismology Department about the aftershocks that are plaguing the New Madrid Seismic Zone. As you heard, no one is willing to forecast when the shaking will end. It does not sound like anyone should plan on that being any time soon."

Chris Nelson sat next to Pamela Weekes on the concrete abutment beside the remains of the Memphis City Building. The two stared out across the remnants of the Pyramid and Mud Island to the swollen

Mississippi River. They kicked their heels against the concrete and relaxed in the sun.

Pamela spoke first. "Chris, I really never expected to see you again. Jenny came by and showed Marion the printout of the last prediction your computer model made. How did you do it? You were right on: time, place, and magnitude."

Chris looked at her, a look of guilt clouding his face. "It didn't say 7.9 when I headed for New Simon."

"But it did the next morning."

Chris shifted uncomfortably. "Pamela, I appreciated our early morning talks and donuts when I was working on my model, but I've given that up. I threw all my work into the river. I've quit seismology."

Pamela looked at him in dismay. "You what?" During the weeks before the earthquake, she had stopped by early in the morning at the Seismology Lab while jogging to visit with Chris, bringing donuts for their 6:00AM meetings. She knew of his work and had been a silent champion for his accomplishments.

"I decided the world doesn't need a prophet. Even if everyone had known the earthquake was going to happen, it still would have done its damage. I felt so much guilt, like I caused the earthquake to happen. I just quit."

Pamela looked down. "I guess I can understand that, but others won't. They'll think you understand the holy grail of seismology Chris. You know they'll come after you, don't you?"

"Yes, but I'll work on that later."

The two sat in silence for a time, lost in their individual thoughts in the peace and quiet. Then shaking from another aftershock coursed through the city, shaking their perch enough to make them uncomfortable.

Pamela broke the silence. "I understand you came down with a couple of other kids. How was the trip?"

Chris brightened. "It was really interesting. We joined an Army convoy coming down Highway-51 and made it into town. Tina was a waitress up at New Simon, right on top of the epicenter. She's still looking for her dad.

"I found my dad. He was killed when his neighbor's house burned. Dad had fixed his house like I told him to and it did fine, but his neighbors never bothered to protect their gas line, and their house blew up when he went to help. Burned him to a crisp.

"Tina's friend Alex came with us, and he's now working on some city project. He used to work for JQ McCrombie and he told me about some of the shortcuts JQ made in some of the retrofit work. From what I hear, most everything JQ built has come down around his ears."

Pamela sighed. "Yes, the last thing we heard was that he must have been killed in his new building when it crumbled."

Chris laughed ironically. "Yeah. I remember seeing you at JQ's open house last Friday night. He gave me and some other people the boot. He was a real bastard."

Pamela shook her head. "I had talked with him earlier in the week and warned him he should have someone look over the reconstruction done on that building. It didn't look very safe to me."

"So his building fell on him. Don't worry. Couldn't have happened to a nicer man." Chris looked up at Pamela. "So now what? Where do we go from here?"

Pamela shrugged her shoulders. "What do you think?"

Chris said, "Now I realize that what I should have been working on is how to recover from such a calamity. I look around and see all this destruction and realize—we don't know what we are doing. Nobody did enough planning and thinking."

Pamela nodded. "You know, you're right. Chris, we're now four days into the rescue phase, which means rescue is effectively over. There just isn't anyone left to rescue. Now we must recover. That will take the next five years or even longer. People just do not understand how bad the situation will be. They have not prepared for recovery. They have no idea what it will entail. The time ahead of us will be the hard part."

Chris looked up. "Can I help?"

Pamela smiled. "Yes. Everyone can help. We have to make up for the lack of thought put into the planning process over the last 20 years. We are faced with rebuilding a whole new city, a whole new river, and maybe even a whole new country."

"Then that's what I'll do. From here on my Holy Grail is recovery."